WETLAND DESIGN

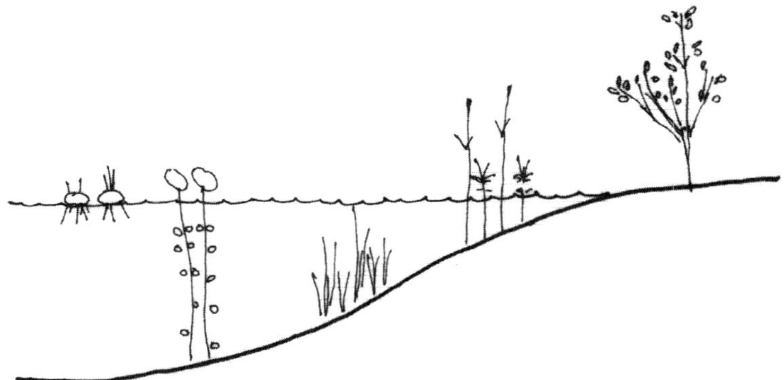

WETLAND DESIGN

Principles and Practices for Landscape Architects and Land-Use Planners

ROBERT L. FRANCE, PH. D.

*ASSOCIATE PROFESSOR OF LANDSCAPE ECOLOGY,
SCIENCE DIRECTOR OF THE CENTER FOR TECHNOLOGY
AND THE ENVIRONMENT (CTE),
HARVARD DESIGN SCHOOL*

ILLUSTRATED BY
CARLOS TORRES AND MATTHEW TUCKER
HARVARD DESIGN SCHOOL

W. W. NORTON
New York • London

To all those who have been captured by the beauty and magic of wetlands, and especially to Henry David Thoreau for his enduring message for modern wetland designers:

"Redeeming a swamp . . . comes pretty near to making a world."

—Henry David Thoreau (*Journal*, IX, 311)

"The turning a swamp into a garden, though the poet may not think it an improvement, is at any rate an enterprise interesting to all men."

—Henry David Thoreau (*Journal*, III, 328)

Printed in the United States of America
First Edition

For information about permission to reproduce selections from this book, write to Permissions, W. W. Norton & Company, Inc., 500 Fifth Avenue, New York, NY 10110

Manufacturing by Courier Westford
Book design by Gilda Hannah
Production manager: Leeann Graham

Library of Congress Cataloging-in-Publication Data

France, R. L. (Robert Lawrence)
Wetland design: principles and practices for landscape architects and land-use planners / Robert L. France; illustrated by Carlos Torres and Matthew Tucker.
p. cm.
Includes bibliographical references (p.) and index.
ISBN 0-393-73073-5
1. Wetland landscape design. I. Title.

SB475.9.W48 F73 2002
712—dc21 2002038681

W. W. Norton & Company, Inc., 500 Fifth Avenue, New York, N.Y. 10110
www.wwnorton.com

W. W. Norton & Company Ltd., Castle House, 75/76 Wells St., London W1T 3QT

0 9 8 7 6 5 4 3 2 1

CONTENTS

ACKNOWLEDGMENTS

Most of the text illustrations were very ably drawn by Carlos Torres, and the maps for site-specific case studies were produced by Matthew Tucker. Mina France aided in manuscript preparation and Helen Whybrow edited the text. I am grateful to the following individuals who provided information about or guided me through the wetland case studies: John Maodayo, Catherine Berris, Jeff Harris, Garret Hollands, Laura Kadlecik, Tom Liptan, Bill MacElroy, Carol Mayer-Reed, Martha Peever, Harm Sloterdijk, Malcolm Whitehead, and Daniel Winterbottom. This work was funded by the Harvard Design School and a Tozier Grant from the Faculty of Arts and Sciences at Harvard University. The principles were fine-tuned through student feedback from several wetlands classes and studios at the Harvard Design School.

INTRODUCTION

Wetlands combine the beauty of aesthetic form and ecological function in a way that few other landforms can match. As such, they have been, and will certainly continue to be, important elements in site design and landscape planning. This book explores the creation, restoration, enhancement, and construction of designed wetlands. It provides a practical guide for wetland design on a local, site-specific scale and also reviews the impact of wetland design projects on watersheds.

Specifically, this book lists and illustrates many key principles in wetland design. It serves to distill a copious, diffuse, and specialized body of scientific and engineering material into an accessible primer, thereby enabling designers to apply these principles within landscape architecture and land-use planning. At the same time, the book serves as an introduction to the wider literature on wetland creation and mitigation. It presents over 150 concepts of wetland design and planning, most in short paragraphs accompanied by simple diagrams.

After a chapter that discusses the history of wetland loss and creation this book explores two categories of wetland design in detail. Chapter Two, Watershed Land-Use Planning, focuses on principles for wetland mitigation replacement in the context of regional landscapes. The practical applications of these principles are examined in several ways. First, I outline general frameworks for the identification, evaluation, and prioritization of potential wetland mitigation sites. Second, I present case studies which incorporate many of the planning principles.

Chapter Three, Site-Specific Landscape Architecture, focuses on principles of wetland design for stormwater management, contaminant treatment, biodiversity sustenance, and ancillary benefits to humans. The practical applications of these principles are examined as well. First, a compilation of plans from the literature illustrates real-world examples of wetland design. Second, I review case studies selected because they incorporate many of the design principles.

It is important to note that the principles in this book are presented as general concepts and should not be construed as rigid instructions to be followed cookbook fashion. Occasionally, design criteria are provided in the form of actual numbers, and these should be considered general estimates, obtained from many different sources, which need to be modified to fit the specifics of a region and the idiosyncrasies of a particular site. The site-specific nature of effective wetland design precludes blind adherence to any detailed instruction manual. In short, these guidelines should be most useful as aids in developing conceptual designs.

This primer will familiarize readers with the terms, concepts, aspirations, and procedures associated with wetland design. With such a familiarity, the landscape architect or land-use planner can better formulate design plans that are cogent, ecologically realistic, environmentally beneficial, and pragmatically operational. She or he can then enter into a constructive dialogue with the specialists on the design team—the hydrologist, biologist, environmental regulator, construction manager, and civil engineer—whose task it will be to implement the design plans. The goal of this book is to improve design concepts by introducing the many possibilities that exist in creating a wetland that is both beautiful and useful. Implicit throughout these pages is the idea that for the well-being of individuals living in increasingly urbanized settings, the form of the created wetland may be just as important as its hydrological, chemical, or biological function.

This book offers an optimistic view of the potential for innovative wetland design. Its ultimate message is that the difference between a single-function, utilitarian wetland and a multipurpose wetland with aesthetic benefits has really much more to do with absence of imagination than it has to do with absence of space. Like Thoreau, we must have large, imaginative dreams when it comes to wetlands: "If there were Druids whose temples were the oak groves, my temple is the swamp," he wrote, as quoted in a poster at the Arcata Wastewater Marsh Interpretive Center, a multipurpose wetland if ever there was one.

NOTE: Both imperial and metric measurements have been supplied throughout the text. However, in the drawings, only the international system has been used.

WETLAND DESIGN

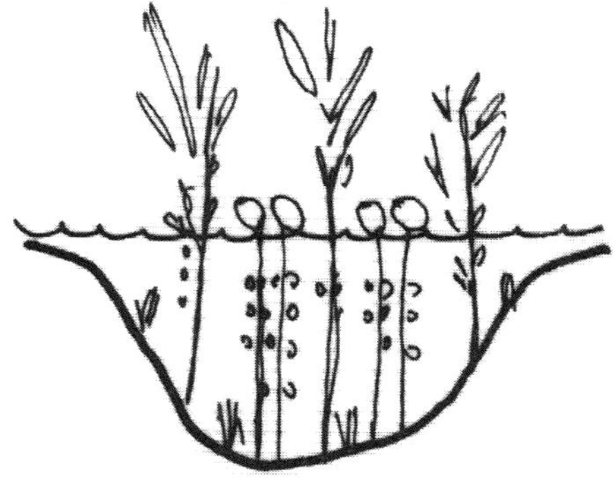

Chapter One

FOUNDATIONS

Historical Markers

Early in the last millennium, a Chinese military commander retired to the old picturesque town of Suzhou. There, by drawing water from one of its famous canals, he created a marvelous garden retreat for emotional and spiritual peace. In his *Chanlang ting* (Pavilion of Blue Waves), the contrast between rocks and water was raised to an art form, with the many wetland plants serving as the focus for the fusion of the real and the illusory in a complex network of small mysterious spaces (1-1). Lotus blossoms waved in the breeze at particular and purposely designed locations where, augmented by "interpretive signage" in the form of quotations from famous poems, their beauty would inspire the host and his guests to write or paint.

1-1

Sometime during the early medieval period, in a remote sequestered valley in Yorkshire, England, an unknown Cistercian monk began conversations with a stonemason about the waterworks system for a future abbey. Together these men, either knowingly or unknowingly, followed a long-established tradition from ancient Rome, China, and Mexico in their incorporation of multiuse hydrologic design. Water from the nearby valley stream was to be diverted into a series of shallow wetlands created primarily for fish aquaculture, but also as an aid for meditative ambiance. The intricately circulated water would then flow underneath the abbey where it would collect and carry the waste from the kitchen and the communal latrines downstream through riparian wetlands before rejoining the valley stream below.

In the mid-nineteenth century, an individual considered to be somewhat odd by his neighbors spent a good deal of his time strolling about the New England countryside, when not writing his iconoclastic, misanthropic, hauntingly beautiful prose. His hydrophilic wanderings drew him time and again to several wetlands (1-2), in which he would enthusiastically paddle, wade, or swim, ever "botanizing" as he went. Henry David Thoreau's lasting legacy, in the form of his copious field notes, which would later find their way into his published *Journal* and other books, expressed as no one had done before—and few have done since—the magic, mystery, and beauty of wetlands as places of both biological fascination and transcendental pantheism.

During the last quarter of the nineteenth century, a brilliant visionary turned his talents toward addressing a major environmental problem that had developed in Boston during a period of rapid urban growth. His

1-2

1-3

plans focused on restoring a precolonial saltmarsh, mitigating potentially dangerous seasonal flooding, and alleviating health risks from sewage discharges, while at the same time providing recreation opportunities and linking the historic city center with its new outlying neighborhoods. Frederick Law Olmsted's work on the Backbay Fens of the Muddy River, part of his larger "Emerald Necklace," were to culminate in what is arguably the world's most famous and beautiful wetland created for urban water treatment, as well as the oldest park system in the United States (1-3).

Near the turn of the twentieth century, an eclectic man, threatened by blindness, began a bold experiment that would become his obsession for the next thirty years. Fascinated by previous visits to the watery world of Venice, this artist, against the wishes of his neighbors and strict zoning laws, diverted a stream to create a wetland over which he built a Japanese bridge. It is doubtful that there is another created wetland anywhere that has so enriched the art world as that of Claude Monet's water lily pond at Giverny. Monet enlarged his wetland three times, its size growing in proportion to his desire to capture light on canvas. His 300-odd water lily pictures represent one of the most beautiful painting cycles in history.

Today the skeletal ruins of the Cistercian monasteries bear mute witness to their turbulent dissolution under Henry VIII, and their fascinating hydrological systems are all but buried. A few of Thoreau's wetland stomping grounds remain temporarily protected from encroaching development by the Walden Woods Project, while the ephemeral wet meadows along the nearby Concord River that were often his haunt have been altered to boost waterfowl abundance through permanent ponding. Olmsted's Fenway has become a dysfunctional and polluted mess due to major hydrological reconfiguration by city planners and engineers which resulted in drainage problems, sediment buildup, and changes to the system's vegetation and overall ecology. And the water gardens of Monet at Giverny and those at Suzhou, both lovingly restored several decades ago, need careful attention to prevent them becoming "loved to death" by hordes of tourists.

Wetland Loss

Wetlands occupy over 3 million sq. mi. (8 million sq. km.) worldwide. In North America, about half of all waterfowl nest in wetlands, and two-thirds of the catch from commercial shellfish and sportfish fisheries derive from wetlands. Over 50 million people are thought to annually observe and photograph wildlife residing in wetlands. That is the good news.

The bad news is that throughout history, people have mistreated and manipulated wetlands to either create harbors and provide homesites and farmland, or simply because they envisioned wetlands as evil dens harboring pests and criminals. Attitudes have changed at a glacier pace. Shockingly, W. Patrick could write as recently as 1994 that:

> A negative view of wetlands has persisted to recent times. About 15 years ago, I proposed a research project to a federal agency to study the fate of plant nutrients that move from uplands to adjacent wetlands. The project was funded, but the federal scientist who was my project officer had changed the word wetlands to wastelands in his description of the work. He had never encountered the word wetland before and thought I had misspelled wasteland. This same federal agency now supports research on wetlands. In Great Britain, the term wetland was not even in use as recently as a decade ago.

The net result of such an attitude has been a loss of about half of the original 197 million acres (80 million ha.) of wetlands in the United States. The figure can be as high as 90 percent loss for certain individual states.

The vast majority of wetland losses result from drainage and clearing for agriculture. Urban development accounts for less than 10 percent of the area of freshwater wetlands that have disappeared. Specific agents of loss include cattle grazing, alteration of chemistry (shifting of nutrient regimes, introduction of toxicants), recreational overuse or negligent abuse, and most important, physical modifications (filling, draining, excavating, clearing, water diversion, flooding, sediment trapping, shading, riparian effects, and so forth).

Cumulative Landscape Impact

A strange imbalance exists between the scale at which wetland losses are felt by society and the scale at which wetlands are preserved. Often the most important role that wetlands play is unrelated to any specific single wetland, but rather to the cumulative effect of many wetlands on large-scale landscape processes. For example, it is no accident that those regions in the United States that have experienced extensive flooding in the last decade—California from El Niño and the Midwest from the Mississippi—are also those regions which have lost over 80 percent of their original wetlands. Nevertheless, permits are still issued on a project-specific, case-by-case basis, leading some to fear that the environment is in serious danger of being "nickel and dimed" to death.

Wetlands are not isolated entities. They are linked to their surrounding landscape in terms of sensitivity to terrestrial disturbance, and they are dependent on the presence and proximity of surrounding water bodies. There is a functional interdependency of wetlands, and since from a watershed perspective they operate together as complexes, their losses are also felt collectively. For example, given that wetlands can be looked upon as habitat islands in a terrestrial sea, their fragmentation into isolated patches can have profound implications on mobile species that require several adjacent wetlands to sustain their life cycles. No wonder then that waterfowl species richness is lower per unit wetland size for regions with isolated wetlands compared to regions maintaining expansive wetland complexes.

Because humans are altering wetlands on progressively larger landscape scales, the consequences of wetland losses can only be properly assessed through regional watershed-based planning.

Replacement Mitigation

Because of the recent recognition of the numerous benefits to people associated with wetlands, the real estate industry is increasingly aware of the economic benefits of residential developments situated near wetlands, to such an extent that such lots bring higher property premiums. What it means is that natural wetlands are being increasingly threatened. What this also means, however, is that there are economic incentives as well as ecological ones to create new wetlands for compensatory loss mitigation.

Mitigation is the procedure whereby the loss of wetlands through (presumably unavoidable) development pressure is compensated for by their replacement elsewhere. This procedure can take the form of direct "in-kind" or "one-for-one" functional creation of new wetlands, or the complete restoration or functional enhancement of existing wetlands.

Depending on jurisdictional regulations, replacement wetlands either must occur as close as possible to the site of the original lost wetland or be located off-site, occasionally agglomerated into spatially large mitigation "banks." The concept of wetland mitigation banks in particular, and even wetland mitigation in general, remains controversial. Proponents of banking stress that its premier attraction lies in consolidating the creation of numerous small, isolated wetlands into one large hydrologically and ecologically favorable site that is easier to monitor and manage. It is thought that the creation or restoration of banks of carefully designed wetlands may be more beneficial to the watershed than those same wetlands created piecemeal throughout the watershed on individual development sites.

Unfortunately, the track record for wetland mitigation is poor. Indeed, most mitigation projects fail because of inadequate design, bad site selection, absence of long-term maintenance plans, and lack of knowledge about the multifaceted roles of wetlands in a watershed context.

A great opportunity now exists for much needed improvement in the field of wetland mitigation, and this is being led by land-use planners educated in the emerging discipline of watershed management. The new emphasis is on mitigating and regulating development impacts on wetlands based on a regional framework of both ecological and sociological relevance.

Wetland Creation

The modern process of formally creating wetlands to function as centers for wastewater or contaminant treatment is only several decades old. Surprisingly, it took engineers and scientists quite some time to realize that many of the biological breakdown processes that they were endeavoring to construct within the controlled environments of sewage treatment plants mimicked processes that already existed naturally in many wetlands.

Over the last decade there also has been increasing recognition of the many added benefits of using created wetlands for either waste- or stormwater treatment. In short, such systems are generally self-sustaining, adapt well to variable contaminant loading rates and water flows, are flexible enough to fit into many diverse landscapes, can certainly be aesthetically attractive, and are almost always relatively inexpensive to build. Today, over one thousand treatment wetlands have been created in North America and Europe.

Created treatment wetlands typically cost about $5,000–$6,000 per acre ($12,000–$66,000 per ha. [mean = $24,000]) to build, an investment that is between 10–50 percent of the costs of conventional chemical treatment facilities, with only about half the costs associated with long-term operation and maintenance. Other estimates suggest that decentralized wastewater treatment systems (such as wetlands) for individual or clusters of five homes can be up to three times less expensive than centralized treatment systems.

Unfortunately, however, many of the treatment wetlands that have been constructed are, although functionally efficient in removing contaminants, most often embarrassingly ugly in their final form. This results from the utilitarian view of nature held by many engineers, whose goal is often to construct a cost-effective functional treatment system, without aesthetic considerations.

A great opportunity now exists to reshape and positively advance the field of treatment wetlands through the field of landscape architecture, whose design motivations encompass a more pluralistic view of the natural environment fostered by a close emotional connection to the created result. This represents a conscious shift in focus from "constructing" wetlands to "creating" wetlands. To borrow Charles Dickens' wonderful lines from *Pickwick Papers* (also quoted in David Salvesen's seminal book on mitigation wetlands): "The whole difference between construction and creation is exactly this: that a thing constructed can only be loved after it is constructed; but a thing created is loved before it exists." This serves as a powerful maxim by which to redirect wetland creation as an evolving and maturing discipline.

Chapter Two

WATERSHED LAND-USE PLANNING

The earth's hydrological cycle has been altered severely through centuries of environmentally deleterious human activity. Due to its corporeal form, water is an important integrator of environmental disturbance across spatial scales ranging from small subdivisions to regional drainage basins. As a result, local, site-specific actions have a cumulative effect on larger landscapes. Given the decreases in environmental integrity that already exist, proactive design projects can provide positive environmental benefits. Wetland creation is one important tool that can offer region-wide landscape benefits in addition to site-specific advantages. This land-use planning section deals exclusively with freshwater wetlands.

Primary Principles

Hydrology Modifiers

Wetlands are "wet lands," and as such are tightly coupled with the hydrologic cycle.

FLOOD REDUCTION

(a) Site-specific. Wetlands operate like giant sponges in that they slow down and absorb excess stormwater runoff, then gradually release the stored water over a prolonged period. This reduces peak flows downstream and lessens the chances of flooding (2-1).

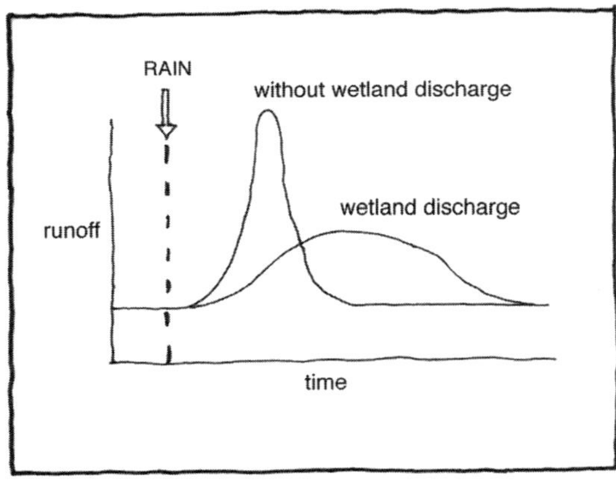

2-1

b) Landscape influence. A strong positive relationship exists between the percentage of upstream wetlands that have been lost and the percentage increase in watershed peak flow discharges (2-2).

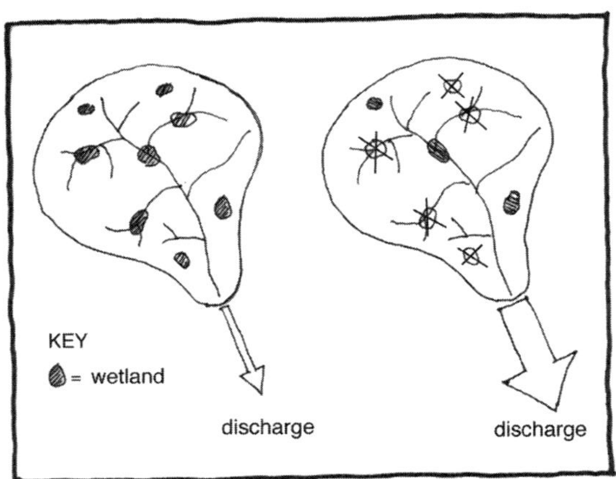

2-2

SEDIMENT CONTROL

Wetlands reduce the velocity of floodwaters, thereby enabling sediments to be precipitated. This reduces downstream deposition as well as further erosive scouring.

Contaminant Sinks

The retention and removal of contaminants in wetlands depends on a complex and inter-related system of chemical transformations involving three major pathways—physical, chemical and biological. Each of these removal mechanisms must be considered in the design of treatment wetlands.

REMOVAL MECHANISMS

(a) Physical. As water moves through a wetland, contaminants are removed through simple physical processes of sedimentation, filtration, adsorption and volatilization.

(b) Chemical. Contaminant removal is simultaneously augmented through chemical precipitation and adsorption in addition to a suite of chemical breakdown reactions (2-3).

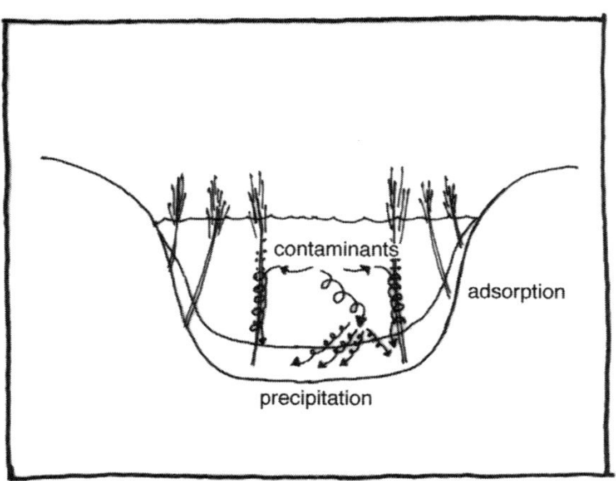

2-3

(c) Biological. Further removal of contaminants occurs via microbial and plant (including algal) metabolism, plant absorption and eventual natural die-off, and sediment accumulation of organic matter (2-4).

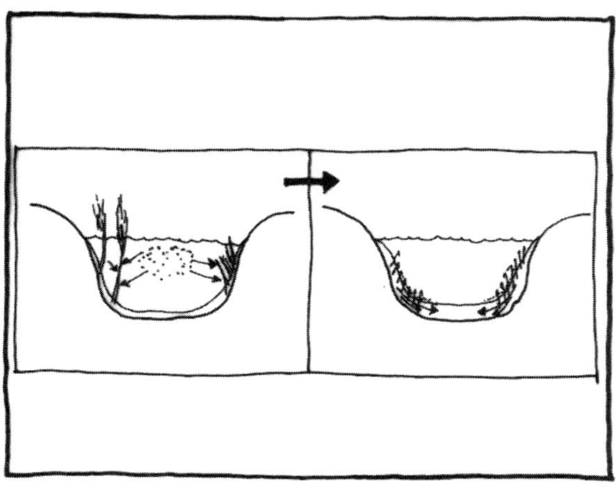

2-4

REMOVAL SPECIFICITY

Removal mechanisms vary depending on the contaminants in question, which may affect overall wetland design by highlighting some beneficial processes of removal over others.

2-11

MANAGEMENT IMPLICATIONS

Newly created wetlands are rapidly colonized by a diverse array of plants and animals. The resulting high levels of abundance mean that these new wetlands can quickly become important natural resources (2-11).

Human Amenities

Wetlands, long regarded pejoratively as "wastelands," are now recognized as providers of a great number of benefits to humans, in addition to their other important roles in hydrology, water quality, and biodiversity.

2-12

PSYCHOLOGICAL SUPPORT

(a) Aesthetics. Wetlands are among the most beautiful of all landscapes. Their great diversity in terms of location, setting, size, shape, and life-form composition provides valued visual enjoyment (2-12).

2-13

(b) Wild open space. Often being situated between residential and industrial areas, wetlands provide a wild buffer to ameliorate the stresses of increasingly urbanized lifestyles (2-13).

RECREATION

Wetlands are extremely popular destinations for land-based and aquatic recreational activities such as walking, jogging, bird watching, angling, photography, painting, boating, and game shooting (2-14).

2-14

HISTORY

Because early natives and colonists often settled close to wetlands due to their role as food sources, these regions may harbor a rich cultural or archeological heritage (2-15).

2-15

EDUCATION

Wetlands offer a myriad of opportunities for formalized natural history study through use of interpretive signage, buildings and kiosks, and instructional programs (2-16).

2-16

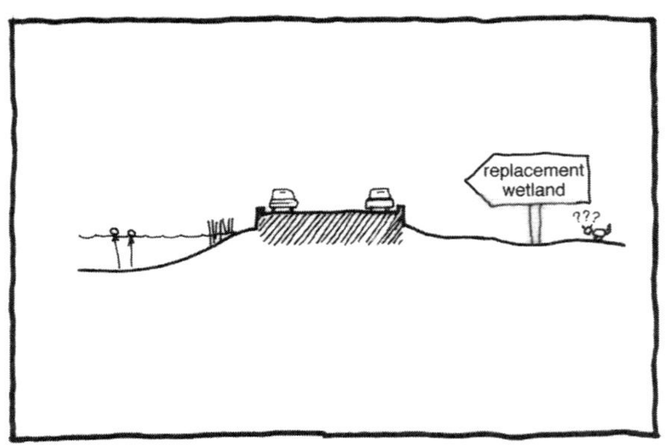

2-21

(b) ***Additive effects.*** Unless checked and closely regulated, replacement mitigation has the potential to significantly alter the type and spatial distribution of wetland ecosystems over vast geographic regions. It is imperative to recognize the cumulative impact of numerous mitigation projects on the entire landscape (2-21).

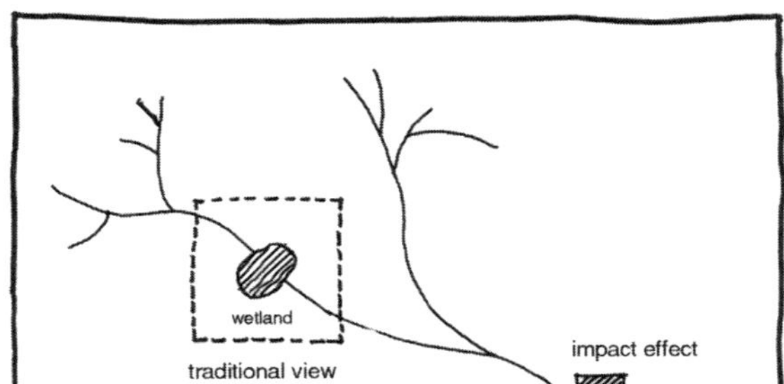

2-22

(c) ***Off-site influences.*** Sometimes the most important environmental repercussions of wetland creation projects occur downstream in the watershed. It is therefore important to look beyond site-specific boundaries (2-22).

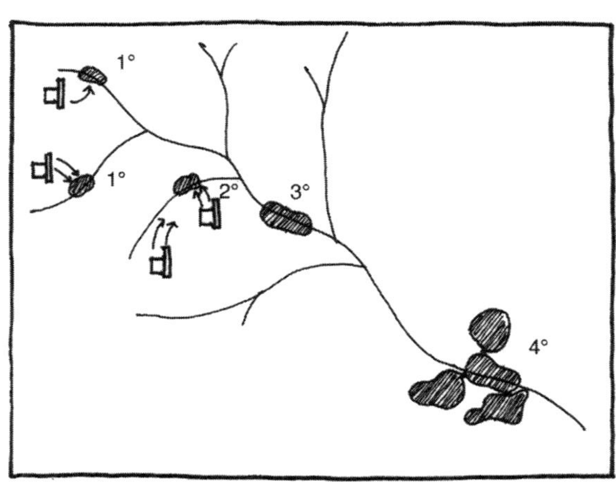

2-23

(d) ***Hierarchical utility.*** The location of a created wetland in a watershed influences its intended role. First-order wetlands involve direct treatment of contaminants at the site; second-order wetlands occur away from the point sources and treat less contaminated water from a variety of sources, but also provide some ancillary benefits; third-order wetlands are often located in riparian zones in the upper watershed; and fourth-order wetlands occur in larger complexes lower in the watershed, and, like third-order wetlands, are as important for flood control and wildlife habitat as they are for contaminant removal (2-23).

PROACTIVE APPROACH

It is essential to begin to identify in advance those wetland sites that will provide the greatest ecological gain to the watershed following restoration. Such an approach is beneficial in that it helps regulators and developers avoid unnecessary and costly conflicts, reduces government regulation, fosters public awareness, allows for assessment of cumulative impacts, and provides an objective framework by which to judge a project's success (2-24).

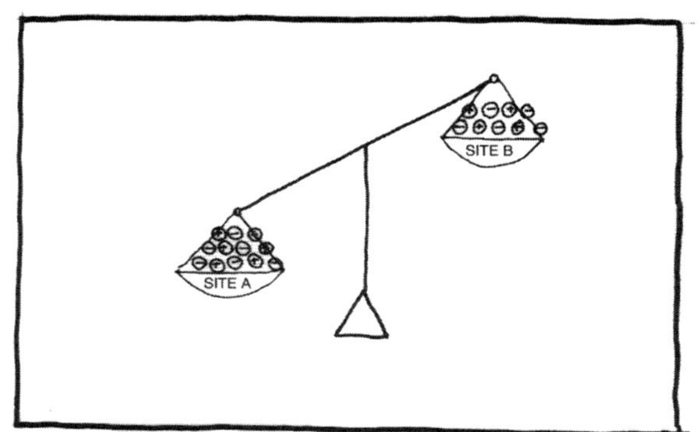

2-24

NATURE KNOWS BEST

The self-designing capacity of nature remains the premier player in ecosystem development, one from which humans can learn a good deal. For example, it can often be preferable to allow native species to recolonize an area rather than to plant predetermined nursery stock. A simple motto for wetland mitigation in land-use planning might be "design less, understand more." Plan for self-design, self-regulation, and self-maintenance.

Initial Policy

The following issues, important for establishing the rationale for wetland mitigation, must be addressed early in the planning process.

TEAM MAKEUP

The need for, and type of, outside consultation has to be recognized and addressed early in the planning stage.

PERMITTING

To limit wasted time, effort, and expenditure, a good understanding of the many regulatory nuances and their appropriate sequencing in the Byzantine legal process regarding wetlands must be recognized, achieved, and applied (2-25).

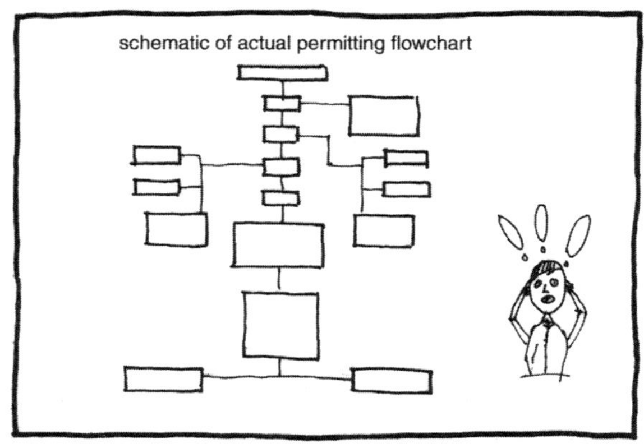

2-25

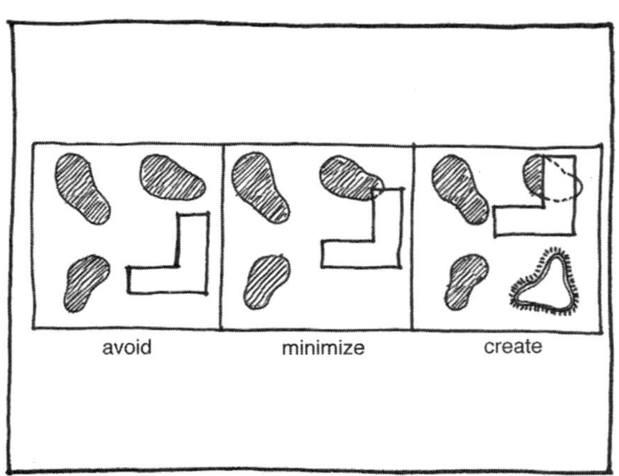

2-26

PROJECT JUSTIFICATION

A logical progression of mitigation justification should be adopted. First, the group must prove that avoidance of threatened wetlands is impossible, then demonstrate how they will minimize development impacts. Finally there must be unequivocal acceptance of the need for creating replacement wetlands as the last resort (2-26).

RESTORATION APPROACHES

At least four strategies for restoration are possible, none of which are mutually exclusive: reestablishing or managing hydrology, restoration of substrata, reestablishing and maintaining native biota, and eliminating or controlling invasive exotics.

Data Acquisition

An inventory of wetland resources is needed to assess their value to the community and the ecosystem, as well as to help establish precise goals and recognizable steps for successful watershed management. This process ultimately results in advance identification of wetlands that may or may not be suitable for filling, helps avoid impact on treasured wetland resources, provides objectives for land acquisition or preservation, and prioritizes potential sites for mitigation.

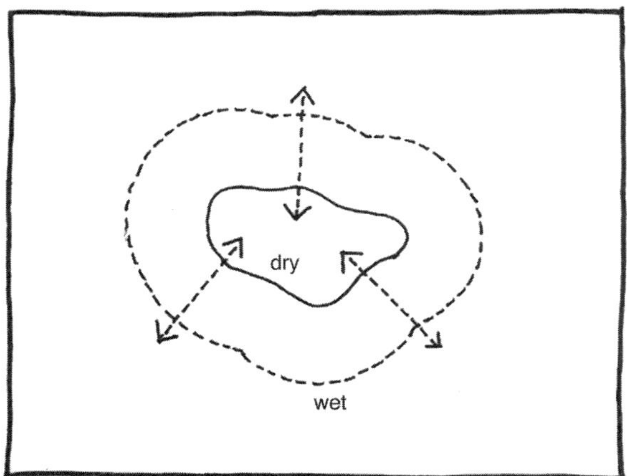

2-27

TARGET WETLAND

(a) Delineation. Although rigid, identifiable borders are required for regulatory purposes, it is important to recognize that wetland boundaries are dynamic entities, fluctuating both seasonally and annually in response to climate and ensuing hydrology (2-27).

(b) Ecological assessment. Within the established wetland bounds, a detailed site-specific survey is needed to determine both static (plant biomass, species identification and diversity, etc.) and dynamic (plant productivity, water and chemical budgets, etc.) variables in order to formally characterize the environment to be lost and replaced.

WATERSHED ANALYSIS

(a) Hydrology. Wetland mitigation feasibility in terms of hydrologic functioning is assessed by broad surveys of water movement patterns in wetlands from source "donors" through "conveyors" to sink "receptors" within the watershed. Knowledge of flood storage capacity, water level fluctuations, and groundwater recharge relationships for all wetlands should be mapped to facilitate achievement of planning objectives (2-28).

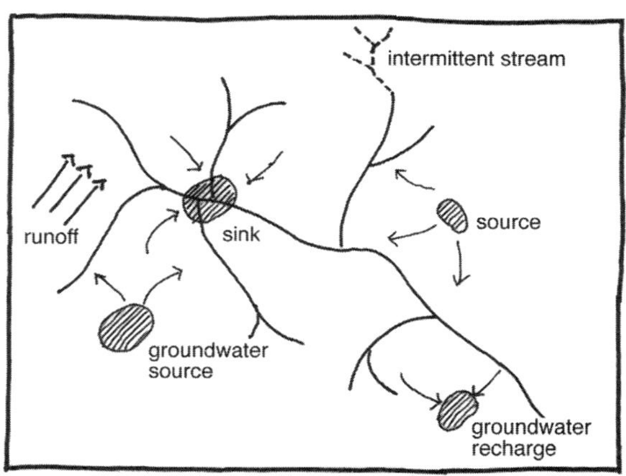

2-28

(b) Chemistry. Systematic sample collection at established hydrograph monitoring sites throughout the watershed is needed to develop comprehensive chemical budgets. Mapping of such information enables calculations of mass loading rates for such elements as nitrogen, if coastally situated, or phosphorus, if located inland; total suspended solids; and recognized problematic contaminants which can be linked to specific watershed locations (2-29).

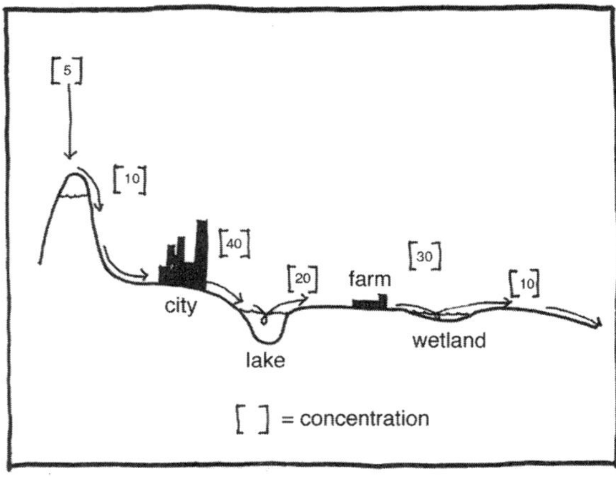

2-29

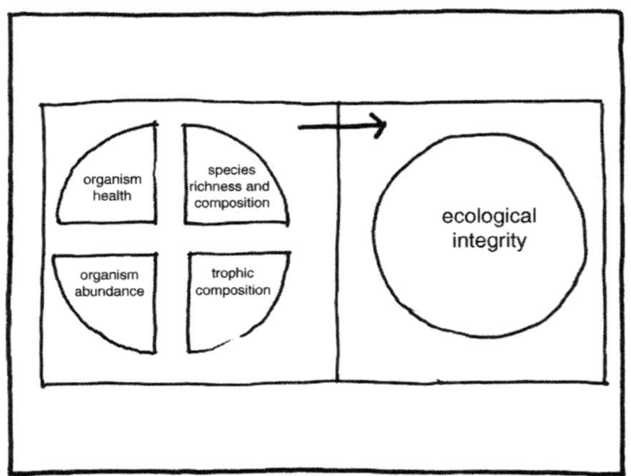

2-30

*(c) **Ecology.*** When the objective is the ability to predict the movement of different groups of animals between habitats based on wetland size, separation distance, and inherent attraction, the subdisciplines of both community and landscape ecology become important in the planning phase. For community ecology, GIS-mapped information concerning species richness, diversity, trophic structure, guild analysis, and biological integrity with respect to assessing wildlife habitat utilization are required (2-30). There is an important need to shift our focus from species-specific approaches to more inclusive, holistic, and functional measures of habitat quality. For landscape ecology, mapped information about the overall mosaic of fragmentation, corridor connectivity, size, shape, edge effects, and juxtaposition patterns with respect to wetland habitat patches are all required. We need to shift our focus from individual wetlands to larger wetland complexes.

*(d) **Human amenities.*** The benefits of retaining wetland fragments in (sub)urban areas for the purposes of education, research, recreation, and aesthetic enjoyment are best assessed in relation to an inventory of similar benefits in other sites in the region. Using a landscape-based analysis of visual quality preferences can help to establish planning standards

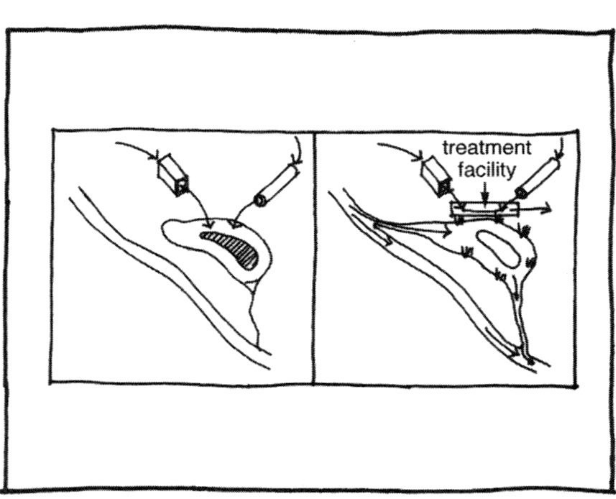

2-31

Mitigation Strategy Adoption

A progression of loss-replacement strategies for mitigation is available depending on the degree of intervention required.

RESTORATION

Restoration rather than creation may involve fewer risks and greater success. Techniques include such procedures as severing problem discharges, removing fill, site regrading, native species replanting, and changing hydrology (2-31).

ENHANCEMENT

Enhancement involves improving existing altered or affected wetlands by strengthening particular functional characteristics, sometimes at the expense of others. In the case of wetlands that are difficult to recreate, often the most feasible mitigation measure is to form, through enhancement, an entirely different type of wetland. For example, a forested or meadow wetland might be changed into an open water area to serve the functions of wildlife habitat, stormwater detention, and beauty (2-32).

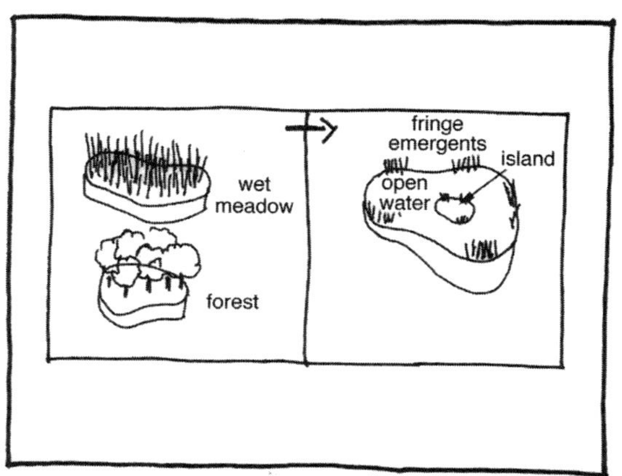

2-32

CREATION

As the last option in wetland mitigation, creation projects offer great opportunities for landscape artistry and innovation which must be considered in association with ecological functionality gains and losses (2-33).

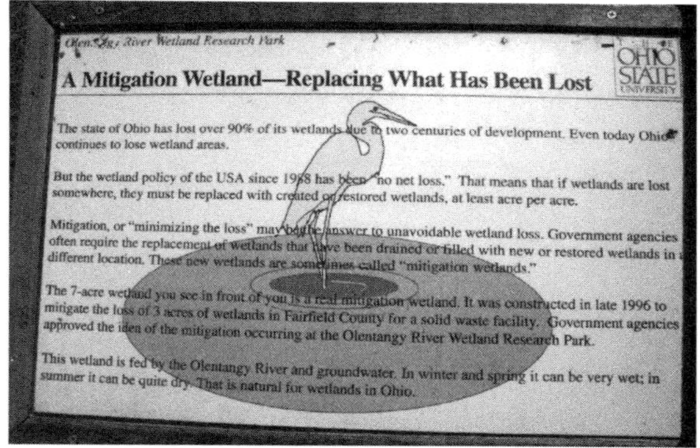

2-33

Site Selection

Screening adjudication of potential mitigation sites involves assessment of wetland functions, project aspirations, and legal possibilities.

FUNCTIONAL EQUIVALENCY EVALUATION

Size is not the only attribute that should be considered in evaluating the health and ecological importance of wetlands that may be lost. Functional attributes such as hydrology, shore stability, nutrient supply, sediment-contaminant retention, groundwater recharge, plant community survival, food production export, and wildlife diversity and productivity are key considerations. These are detailed in the site-specific design guidelines.

LOCATION OPTIONS

Successful wetland mitigation is as dependent on appropriate selection of the replacement site as it is on the actual design of the chosen site. Ideal-

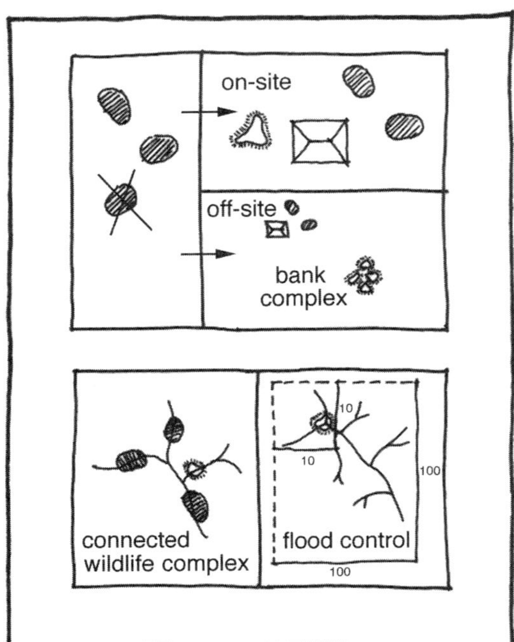

2-34

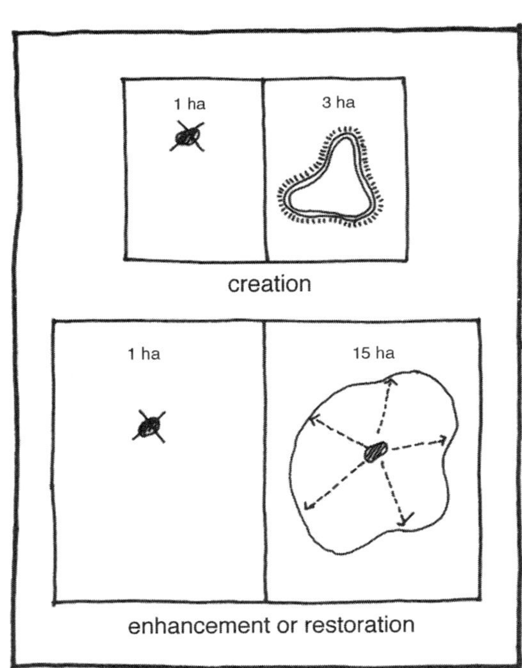

2-35

ly, replacement wetlands should be located as close to the original site as possible in order to maintain watershed functions. However, in certain situations with space requirements and constraints or regulatory possibilities and constraints, offsite mitigation may provide greater cumulative benefits. For example, for replacement wetlands designed for wildlife, primary sites include isolated "oasis" wetlands or those connected in wetland networks rather than those evenly scattered over the landscape (lower panel). And for wetlands designed for flood attenuation, the ideal location is often a position in the watershed where all upstream waters occupy about 10 percent of the surface area (lower panel). Finally, grouping mitigation wetlands into centralized banks, and locating them to provide the greatest potential watershed benefit, is one approach that many consider positive (upper panel) (2-34).

Plan Formulation

Loss-replacement mitigation planning addresses spatial implications of wetland creation projects.

COMPENSATION RATIO DETERMINATION
Compensation replacement ratios must be empirically determined by measuring such wetland functions as wildlife productivity and species diversity or flood attenuation. This will determine the size of the created wetland needed to compensate for the area of wetlands lost during development. Such explicitly stated objectives of compensation ratio targets help developers with advance project planning. In all cases, to accommodate the often poor performance of created wetlands due to unforeseen problems, compensation ratios must be very generous. Typically, at least 2–3 acres (1–2 ha.) are created for every acre lost for direct in-kind mitigation, and up to 4–20 acres (2–8 ha.) are restored or enhanced for every acre lost for restoration or enhancement mitigation (2-35).

MITIGATION PROPOSAL

In addition to components such as detailed descriptions of planting, monitoring and contingency plans, construction methodology, and cost estimates (as subsequently described in the site-specific design guidelines), mitigation proposals must also include formalized descriptions of the area lost and the wetland expected to be created, restored or enhanced.

Follow-up

SUCCESS RECOGNITION

(a) ***What.*** The establishment of a biologically diverse wetland characterized by self-supporting productivity with long-term sustainability is a good marker of mitigation success.

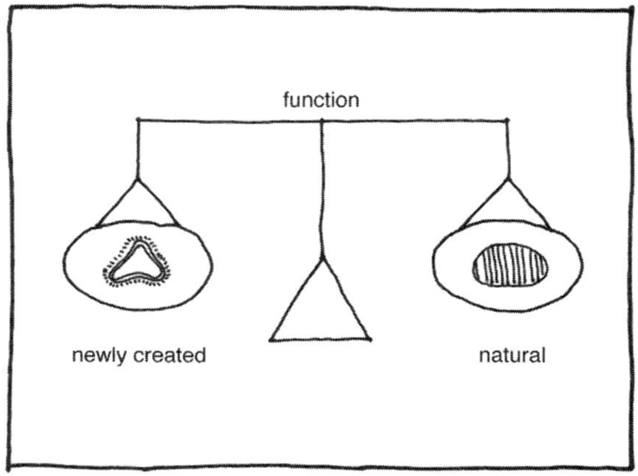

2-36

(b) ***How.*** Evaluation of functional·equivalency with other natural wetlands in the watershed provides the best adjudication of mitigation success (2-36).

(c) ***When.*** Simple measures of 'success' based on once or twice per year monitoring for only 3–5 years are grossly inadequate for true ecological evaluation. Mitigation wetlands must be frequently reassessed, for decades in some cases, before formal judgments can be made of final project success (2-37).

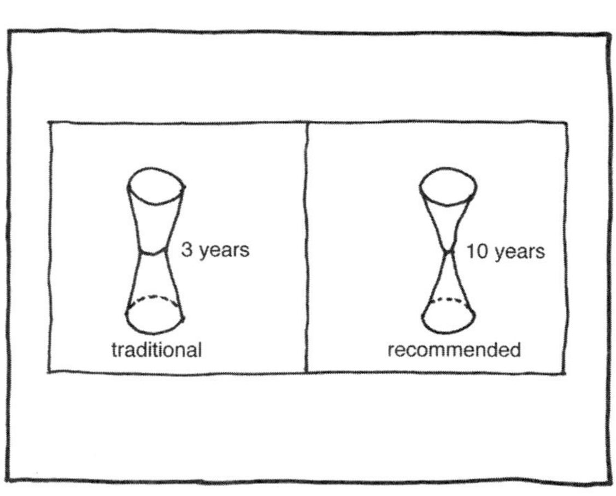

2-37

Practical Applications

General Frameworks for Site Identification and Evaluation

The approaches outlined below represent three of the most heuristically appealing, scientifically detailed, and logistically feasible planning tools for wetland mitigation in watershed management.

Massachusetts Method for Wetland Restoration

Because wetland replacement projects in Massachusetts have experienced a 36 percent failure rate, a watershed rather than a site-by-site approach was warranted to integrate wetland restoration with other watershed management activities. This has led to a science-based framework for identifying potential wetland restoration sites and evaluating their capacity for contributing to watershed improvement in terms of water quality, flood storage, and fish and wildlife habitats. This framework identifies individual wetland restoration sites for compensatory mitigation, including banking and proactive restoration, leading to a comprehensive watershed wetland restoration plan.

Site identification and evaluation may be summarized as follows:

- Identify degraded and former wetlands within the watershed.
- Set watershed goals by identifying functional deficits and opportunities.
- Evaluate degraded wetland sites to assess their functional capacity to meet watershed goals.
- Screen sites against other ecological, logistical, and constraining factors.
- Display results as maps and graphs.

Step 1. Identification of degraded and former wetland sites. Collect background information; delineate watershed boundary; review information on impact patterns; identify potential wetland restoration sites; classify wetlands; identify priority wetland restoration sites.

Step 2. Identification of watershed functional deficits. Identify wildlife habitat goals; identify floodwater storage deficits; identify water quality problems; set restoration goals and objectives.

Step 3. Watershed functional improvement screening questions.
Floodwater storage potential

- Is the site upstream of a flooded area?

- Is the site located in the one-hundred-year floodplain?
- Does the site have or could it be easily modified to create a constructed wetland?
- Are channels absent at the site or, if present, can they be modified?
- Does the site have a grade that produces sheet flow conditions?
- Does or will the restored site support dense woody vegetation?

Water quality enhancement potential

- Is the site in a position in the watershed to improve water quality (for example, located at or downstream from a known area of poor water quality)?
- Is surface water the primary contributor to this site?
- Are the soils at the site conducive to the removal of nutrients?
- Is the site permanently saturated or ephemerally flooded?
- Does the site have a grade that produces sheet flow conditions?
- Is the site restricted such that contaminated surface water is retained and treated by the wetland?
- Does or can the site support dense emergent or woody vegetation?

Wildlife habitat enhancement potential

- Will the site improve the overall quality of the wildlife habitat in the watershed?
- Will the restored site improve the overall diversity of wildlife habitat in the watershed?
- Will the restored site improve the overall connectivity of wildlife habitat in the watershed?

Step 4. Additional site screening questions

- What type of ownership is associated with the site?
- What is the estimated cost of the restoration?
- What type of manipulation is needed to cause restoration?
- What is the level of difficulty in restoring a particular type of wetland?
- What is the degree of surrounding landscape disturbance?
- Are there any exotic or undesirable invasive plant species at or around the site?
- What is the level of difficulty in restoring those functions that were determined to be improved through restoration?
- Are sensitive resource areas for endangered species near the restoration site?
- Are the sites in close proximity to a landfill?
- What is the level of development pressure for each community in the watershed?

REFERENCES

Foote-Smith, C. "Criteria and Procedures for Regional (Watershed) Identification of Wetlands Restoration Sites." *In Wetlands and Watershed Management: Science Applications and Public Policy*, edited by J. A. Kusler., D. E. Willard and H. C. Hull. Berne, NY: Associate State Wetland Managers, 1997: 246–250.

Massachusetts Restoration and Banking Program. *Watershed Wetlands Restoration Planning Guidance*. Boston, MA: Massachusetts Executive Office of Environmental Affairs, 1996.

Massachusetts Restoration and Banking Program. *Site Identification and Evaluation Procedures*. Boston, MA: Massachusetts Executive Office of Environmental Affairs, 1996.

New England U.S. Army Corps of Engineers. *Neponset River Watershed Restoration Analysis: Section 22 Planning Assistance to States Program.* Boston, MA: Massachusetts Executive Office of Environmental Affairs, 1997.

Rapid Assessment of Wetland Functionality

Given that one of the major purposes of effective watershed planning is to characterize the capabilities and limitations of wetland resources, and that hydrology is recognized to be the premier determinant of wetland functionality, establishment of a hydrogeomorphic (HGM) classification system makes sense. With such a system, the manner in which wetlands receive and release water into the landscape provides insight into how they might respond to the watershed effects of urbanization.

The popular HGM classification system, with its emphasis on water source and vulnerability, identifies three major wetland types. Donor wetlands receive water principally from precipitation and deliver it downstream by overland or channelized flow; receptor wetlands receive water from surface discharges of groundwater; and conveyor wetlands transmit water by overland or channelized flow. Each of these wetlands types is susceptible to its own threat from urbanization: donor wetlands are often drained, receptor wetlands lose water through interference with aquifer recharge, and conveyor wetlands are altered through river straightening or constriction and consequent changes in hydraulic loading.

The HGM field assessment is designed to be conducted in conjunction with wetland boundary delineation and is based on simple qualitative appraisals of hydrologic connectivity for each region of investigation:

- Develop a general profile for each HGM class.
- Develop a list of functions.
- Develop a functional profile for each HGM class.
- List the relevant and appropriate variables for each function.
- Describe each of the variables.

- Prepare rationale for model development based on weighted functions.
- Develop an inventory sheet.
- Create a set of functional indices.

The functional indices generated by the models for each HGM class provide an objective means by which to compare and assess the role of wetlands in a watershed context. This process aids in ranking suitable sites for potential restoration or in importance when considering mitigating loss.

REFERENCES

Brinson, M. M. "Changes in the Functioning of Wetlands Along Environmental Gradients". *Wetlands* 13 (1993): 65–74.

Brinson, M. M. "Functional Classification of Wetlands to Facilitate Watershed Planning." In *Wetlands and Watershed Management*: 65–71.

Magee, D. W., and G. G. Hollands. *A Rapid Procedure for Assessing Wetland Functional Capacity Based on Hydrogeomorphic (HGM) Classification*. Berne, NY: Associate State Wetland Managers, 1998.

Evaluation for Planned Wetlands

The egregiously poor performance of wetlands that have been constructed, restored, or enhanced for purposes of mitigation necessitates a more careful appraisal of wetland functional ecology. The Evaluation for Planned Wetlands (EPW) is an approach that rapidly assesses whether a planned wetland has been adequately designed to achieve defined functional goals. The procedure is based on assessing six wetland functions in such a way that the designer can easily obtain information needed to improve the resulting wetland plan. Differences between wetlands are expressed in terms of "Functional Capacity Indices" (FCI) which are dimensionless variables (based on a simple assessment model that combines rated element scores) describing an individual wetland's relative capacity to perform a particular function. These indices are then used to calculate "Functional Capacity Units" (FCU) which serve as a standardized basis for comparing functional differences among wetlands of different size (for example, FCU = Area x FCI).

The six evaluation functions are defined as follows.

- Shoreline bank erosion control: capacity to provide erosion control and to dissipate erosive forces at the shoreline bank.
- Sediment stabilization: capacity to stabilize and retain previously deposited sediments.

- Water quality: capacity to retain and process dissolved or particulate materials to the benefit of downstream surface water quality.
- Wildlife: degree to which a wetland functions as habitat for wildlife as described by habitat complexity.
- Fish: degree to which a wetland habitat meets the food, cover, reproductive, and water quality requirements of fish.
- Uniqueness/heritage: presence of characteristics that distinguish a wetland as unique, rare, or valuable.

The evaluation is conducted as follows:

- Define scope of evaluation: define evaluation objectives; select functions.
- Characterize wetland assessment area: identify project area; delineate assessment area; prepare maps; complete summarizing cover sheet.
- Assess wetland assessment area: complete data sheets on wetland scores; calculate functional capacity indices for each evaluation function; calculate functional capacity units for each evaluation function.
- Set goals: define goals of planned wetland; define type of comparison (baseline or time interval); determine target functional capacity units; estimate minimum area required to meet goals.
- Select planned wetland site: identify and screen potential sites, and select a site in relation to development constraints (based on required site characteristics, planning institutions, project funding, and construction).
- Design planned wetland: identify conditions needed to achieve planned wetland goals (refer to target functional capacity indices and units calculated previously); prepare design (conceptual design, construction plans and specifications).
- Assess planned wetland design: complete data sheets; calculate functional capacity indices and units as described above; determine whether goals are met.

REFERENCES

Bartoldus, C. C., E.W. Garbisch, and M. L. Kraus. *Evaluation for Planned Wetlands (EPW): A Procedure for Assessing Wetland Functions and a Guide to Functional Design*. St. Michaels, MD: Enviormental Concern, Inc., 1994.

Bartoldus, C. C. "EPW: A Procedure for the Functional Assessment of Planned Wetlands." *Water Air Soil Pollution* 77 (1994): 533–541.

Case Studies in Brief

The following ten studies were selected to illustrate a variety of approaches taken in land-use planning. Each approach was motivated by a different mandate addressed in relation to a variety of problems and locations. Six of the studies have been published as scientific papers (each in a different journal), and were chosen for their easy comprehension by a general audience. The other four studies are recent reports of watershed assessment of wetland function made in different regions and using different methodologies.

Landscape Functionality

Flood Storage Capacity (Minnesota, U.S.A.)

Cumulative impacts consider the incremental effect of past, present, and potential future stresses. The cumulative function of all wetlands in a watershed may be different from that of the additive function of individual wetlands themselves. Also, functions that can be locally important may disappear in a downstream context.

This study used geographic information system (GIS) mapping, multiple regression analysis, and regional hydrology equations to estimate the contribution of flow from fifteen watersheds surrounding Minneapolis-St. Paul, Minnesota. Flow rates were then related to measurements of streamwater chemistry to examine water quality concerns.

A nonlinear relationship was found between percent watershed storage of water in wetlands and lakes and the estimated contribution of flow to the one-hundred-year flood potential. Once the storage of water drops to less than 10 percent of the total watershed area, the estimated flood magnitude increases rapidly with further loss of storage. This in turn has serious consequences for the transport of deleterious chemicals in stormwater runoff.

REFERENCE

Johnston, C. A., N. E. Detenbeck, and G. J. Niemi. "The Cumulative Effect of Wetlands on Stream Water Quality and Quantity: A Landscape Approach." *Biogeochemistry* 10 (1990): 105–141.

Sinks and Sources of Phosphorus Loading (Vermont, U.S.A.)

Wetland area alone may not be the best measure of wetland function in a landscape perspective. The location of wetlands in the watershed in terms of their hydrological connectivity to, and potential for chemical coupling with, other surface waters may more effectively capture wetland function.

Lake Champlain, straddling the borders of Vermont, New York, and Quebec, is beginning to eutrophy due to phosphorus inputs from forest clearance and agricultural development. This research project undertook a landscape-limnological (lake) analysis using geographic information system mapping, phosphorus loading estimation, and multivariate (regression) statistics of eight small watersheds on the Vermont side of the lake to determine the movement of phosphorus into streams feeding the lake.

By including only riparian, rather than total wetland area in the regression models, researchers significantly improved their ability to predict phosphorus export. Two-and-a-half acres (1 ha.) of riparian wetland were found to be about thirty-five times more important in phosphorus reduction than 2.5 acres (1 ha.) of agricultural land. This work demonstrated that all wetlands in a watershed do not have the same functional ability to influence chemical behavior. Furthermore, not only are riparian wetlands the most important due to their proximity to streams, wetlands in narrower, low-order streams (rather than larger, high-order streams farther down the watershed) were found to be particularly important for water quality improvement.

REFERENCE

Weller, C. M., M. C. Watzin, and D. Wang. "Role of Wetlands in Reducing Phosphorus Loading to Surface Water in Eight Watersheds in the Lake Champlain Basin." *Environmental Management* 20 (1996): 731–39.

Rural Development and Biodiversity (Ontario, Canada)

Road construction and forest conversion to agricultural land poses a serious threat to wildlife. Existing wetland policies that focus on regulating development within or immediately around wetlands may not provide adequate protection for biodiversity.

This research project used multiple regression analysis to describe relationships between richness of four taxa (birds, mammals, reptiles and amphibians, and plants) and the wetland area, nearby road density, and forest area in thirty southeastern Ontario watersheds.

As expected, strong relationships were observed between wetland area and species richness. In addition, species richness of all taxa except mammals was correlated negatively to the density of paved roads on lands up to 1.2 mi. (2 km.) from the wetland. Reptile/amphibian and mammal richness showed a strong positive correlation with the proportion of forest cover on lands within 1.2 mi. (2 km.) of the wetland. Most interestingly, removal of 20 percent of the forest cover or a 3 ft./2.5 acres (2 mi./ha.) increase in total paved road density on lands within 0.6 mi. (1 km.) of a wetland was predicted to have the same impact on species richness as the loss of 50 percent of the wetland itself. Therefore, land-use practices around wetlands may be as important as the actual size of wetlands in terms of preserving biodiversity.

REFERENCE

Findlay, C. S., and J. Houlahan. "Anthropogenic Correlates of Species Richness in Southeastern Ontario Wetlands." *Conservation Biology* 11 (1997): 1000–1009.

Comprehensive Spatial Survey

Characterization and Prioritization (Colorado, U.S.A.)

An advance identification program was implemented in San Miguel County, Colorado, to provide information to facilitate informed decision-making regarding development in and near wetlands. About 3 percent of the county's 126,716 acres (51,280 ha.) is occupied by wetlands, one quarter of which occur in complexes of two or three different wetland types. Three impacts were assessed: direct effects—the loss of wetland area or functional capacity; indirect effects—changes in wetland type or water quality; elimination of riparian buffers; reduction in functional ability; and cumulative effects—all combined.

To evaluate the functional performance of wetlands, the following hydrological, chemical, biological, and aesthetic information was obtained: data on groundwater recharge and discharge, flood storage, shoreline anchoring, sediment trapping, long- and short-term nutrient retention, downstream fish and wildlife food chain support and habitat, and passive and active recreation.

The ratings for food chain support depend on the direct or indirect use of nutrients, in any form, by animals inhabiting aquatic environments either within a wetland basin or downstream. For downstream sites, food chain indicators are presence of an outlet, non-acidic waters, non-sandy substrate, non-permanent flooding, dense and diverse vegetation with high sustained productivity, non-stagnant or hypersaline water, good flushing rate, and overhanging riparian vegetation. For within-basin sites, good indicators are presence of non-stagnant water, highly productive vegetation, irregular shape with no outlet, good mixing of water, and areas that are neither solely shallow nor unduly warm in the summer.

The ratings for wildlife habitat function depend on the presence of components important to biological integrity, such as structural diversity, openwater patches, wetland size and proximity to nearby waterbodies, and riparian vegetation. Specific variables include those physical and chemical factors that affect the metabolism and refuge of adult and larval fish, as well as the food and cover needs of wildlife species. For fish, these attributes are presence of deep openwater, non-acidic, turbid or flashy water, no barriers to migration, no oxygen stagnation, no artificial water level fluctuations, mesotrophic (medium) productivity, and cool water with riparian shade. For wildlife, these attributes are presence of a good edge ratio, islands, high plant diversity, alkalinity, large and sinuous and irregular basin shape, gentle gradient, no artificial water level fluctuations, small amounts of moss, some openwater, abundant food sources, and distance from human influences.

The wetlands that registered the highest overall values (not only for wildlife but also for the other functions) for prioritizing their preservation included those located along streams and those occurring in large complexes.

REFERENCES

Cooper, D. J., and D. Gilbert. *An Ecological Characterization and Functional Evaluation of Wetlands in the Telluride Region of Colorado*. Boulder, CO: Environmental Protection Agency, 1990.

Science Applications International Corporation. *Ecological Characterization of Wetlands in Eastern San Miguel County, Colorado*. Contract No. C4-68-0072. Denver, CO: Environmental Protection Agency, 1998.

Multiple Objectives in Wetland Management (Oregon, U.S.A.)

The unexpected discovery of wetlands in a major industrial area of Eugene, Oregon, motivated a detailed study of resource assessment and development pressures. In particular, the role of wetlands in flood control, wildlife habitat, stormwater quality improvement, recreation and education, and scientific research was investigated. It was feared that the piecemeal permit process for individual sites was having a cumulative fragmenting impact on the landscape, so an integrative planning procedure was implemented.

The overall wetland plan objectives were to educate the public about potential choices available; find a balance between protection and development; be inclusive of all interests in the community; and turn what may have been perceived as a development "problem" into a beneficial opportunity.

Much effort was spent on developing a framework for translating wetland science to policy decisions. The strategy was to move from wetland classification (based on characterizing ecological functions) toward a ranking or categorization of wetland values by use of a process of citizen review and involvement.

Specifically, the planning process began with a broad framework in which goals and visions were established. Then, following the acquisition and assessment of functionality data, a set of alternatives were clearly articulated to the public that balanced functions, values, and land uses. The resulting implementation plan included zoning, permitting, land purchase, restoration, mitigation banking, and instructions for management (including monitoring and financing).

The final categorization process led to four designations on the produced map: wetlands to be protected; wetlands to be restored; wetlands whose loss from development would not invoke serious consequences; and upland areas to be protected to ensure wetland and riparian connectivity.

REFERENCES

Gordon, S. C. "West Eugene Wetlands Program: A Case Study in Multiple Objective Water Resources Management Planning." In *Wetlands and Watershed Management*: 119–136.

Proactive Restoration (Massachusetts, U.S.A.)

Because the Neponset River Watershed, which covers 115 sq. mi. in metropolitan Boston, falls less than 300 ft. over its 28-mi. length, wetlands were once prevalent along its banks. Substantial losses have since been attributed to ditching and diking for mosquito control and agricultural purposes, earth moving, fill for roads and buildings, gravel pits, sedimentation, excavation, streambank clearing, golf courses, impounding, channelization, forest cutting, railroad embankments, and power line right-of-way maintenance. Water quantity and quality problems include flooding damage, elevated temperatures and heavy metals, decreased oxygen, increased sedimentation and turbidity, nutrient and bacterial pollution, and invasive plants. Urban development and habitat fragmentation have resulted in declines in the quality of fisheries and the number of rare species.

The project goal was to develop and implement a watershed-wide restoration plan that would improve the overall health of the watershed by returning some of the lost wetlands. Specifically, wetland restoration goals were improving water quality, restoring salt marshes, improving wildlife habitat, improving flood storage, addressing problems related to invasive species, improving coldwater fisheries habitat, and improving groundwater recharge and stream baseflow.

The approach evaluated 171 potential restoration sites (ranging in size from 1–250 acres [0.25–100 ha.]) identified by geographic information system (GIS) and hydric (waterlogged) soils mapping, based on their capacity to address documented wetland problems. Specifically, sites were prioritized through an evaluation process designed to highlight the capability of each site, once restored, to improve the overall watershed function in terms of flood water storage, water quality, and fish and wildlife habitat.

Of the original screened sites, sixty-five were identified as being priority locations for contributing to the varied watershed-level restoration goals. Specifically, those sites with the greatest potential to provide significant water quality improvement were located downstream of developed areas or agricultural lands where sediment, nutrients or heavy metals were likely to enter and accumulate. Sites with the greatest potential to mitigate flood damage were located upstream of threatened floodplain areas. Of the four major wetland complexes identified, cedar swamps and estuarine wetlands were recognized as potential restoration sites for enhancing wildlife diversity either through enlargement or by increasing inter-wetland connections.

REFERENCES

Massachusetts Restoration and Banking Program. *Neponset River Watershed Restoration Plan: Preliminary Report.* Boston, MA: Massachusetts Executive Office of Environmental Affairs, 1997.

Massachusetts Restoration and Banking Program. *Neponset River Watershed Restoration Plan: Executive Summary.* Boston, MA: Massachusetts Executive Office of Environmental Affairs, 1998.

New England U.S. Army Corps of Engineers. *Neponset River Watershed Restoration Analysis.*

Creation Projects

Agricultural Runoff Treatment (Maine, U.S.A.)

Water pollution from non-point sources is often the most significant and difficult to control. Effective management of such watershed problems entails a combination of survey, mitigation, and modeling procedures.

Long Lake, Maine, has a high recreation potential, experiences substantial shoreline development, and is surrounded by intensive agricultural land use (potato farming). Survey data and modeling determined runoff from potato farms to be the main source of sediment and phosphorus in the watershed.

In this study, nutrient/sediment control systems (a sedimentation basin, grass filter strip, wetland, and detention pond in series) were constructed to treat runoff from targeted problem areas. The mean efficiency of 75 percent for phosphorus removal determined from one such treatment system was scaled up to the landscape level in order to model what the effects of installing an additional nineteen systems might be. It was estimated that phosphorus loading to the lake would be reduced between 10 and 33 percent under this management plan when implemented in association with the diversion of municipal wastewater, the major point source of pollution in the watershed. Wetland-pond systems can therefore be a very cost-effective means to treat non-point source pollution on a watershed basis.

REFERENCE

Bouchard, R., M. Higgins, and C. Rock. "Using Constructed Wetland-pond Systems to Treat Agricultural Runoff: A Watershed Perspective." *Lake and Reservoir Management* 13 (1996): 29–36.

Regional Nitrogen Export Model (Sweden)

Due to increased atmospheric deposition and riverine loading, the nitrogen concentration in the Baltic Sea has increased. This has necessitated international agreements to reduce terrestrial inputs by up to 50 percent. One way to bring about such a reduction is through the creation of wetlands with known capabilities to retain nitrogen. However, the wetland retention efficiency obtained from studies of individual wetlands in isolation cannot be extrapolated in a simple linear fashion to estimate the net reduction of nitrogen transport at the mouth of an entire watershed.

This research project in southern Sweden developed an accurate regional model of nitrogen transport in association with detailed measurements of nitrogen retention at selected representative wetlands. The degree of nitrogen retention resulting from the creation and placement of hypothetical wetlands throughout the watershed was examined through a series of different modeling scenarios.

It was predicted that the conversion of 1 percent of the watershed into new wetlands would reduce nitrogen transport by 10–16 percent. More than 5 percent of the watershed would need to be converted to halve the amount of nitrogen transported. Creating wetlands in the main stream channels below lakes was estimated to be more effective in reducing transport (and less costly) than creating wetlands at other watershed locations.

REFERENCE

Arheimer, B., and H. B. Wittgren. "Modeling the Effects of Wetlands on Regional Nitrogen Transport." *Ambio* 23 (1995): 125–129.

Highway Construction (Pennsylvania, U.S.A.)

Despite an initial public consultation process to identify and incorporate concerns of different interest groups into project planning, design, and construction, controversies can still arise. In some circumstances, the type of wetland to be replaced deserves as much attention as does the location where the projected mitigation will proceed.

In south-central Pennsylvania, 119 small groundwater seepage wetlands (55 percent classified as forested, 25 percent scrub/shrub, 20 percent emergent and 1 percent open water) totaling 38 acres (15 ha.) were disturbed by the construction of a new highway. A detailed process involving groundwater and water budget analysis, soil characterization, land use availability, archeological considerations, wildlife evaluation, and excavation costs was used in a hierarchical prioritization to identify potential replacement sites. Five wetland sites totaling 55 acres (22 ha.) were selected to replace the functions and values of the lost wetlands, and another 150 acres (60 ha.) of buffer zone areas were purchased to provide additional habitat for wildlife. Fifty habitat enhancement structures were installed to further attract wildlife. The project was believed to be a success because of the wide variety of wildlife occupying the replacement wetlands, which are very popular for school groups and birdwatching clubs.

For some wetland purists, however, this award-winning replacement scheme was unsatisfactory. Because the new wetlands were not headwater seepage in origin, but fed by surface water and located in the floodplain, they, strictly speaking, did not recapture the precise functions of the lost wetlands. Furthermore, the new wetlands contained perennial openwater rather than periodically drying out as the original natural floodplain wetlands had done. These critics contend that the abundant wildlife, while photogenic, is not typical of natural floodplain wetlands, and that because groundwater seepage wetlands cannot be accurately duplicated, impacts such as the highway construction project produced should be avoided. Also, the strategy of mitigation banking may be unrealistic to apply to locations where the majority of the natural wetlands are small and scattered about the landscape rather than concentrated into a few areas.

REFERENCES

Cole, C. A., R. P. Brooks, and D. H. Wardrop. "Building a Better Wetland: A Response to Linda Zug." *Wetland Journal* 10(2) (1998): 8–11.

Zug, L.S. "Habitat, Water Quality, and Wetland Preservation: U.S. Route 220 (I-99) Replacement Wetlands, Blair County, Pennsylvania." *Wetland Journal* 9(4) (1997): 3–7.

Water Quality Improvement (West Lake, China)

The city of Hangzhou, located near Shanghai, was probably the largest city in the world when Marco Polo visited it during the thirteenth century. The city borders West Lake, one of the most visited sites in all of China due to its stunning beauty and historic importance as the former capital. However, despite centuries of poems attesting to its scenic charms, West Lake is currently experiencing extreme eutrophication due in part to high nutrient inputs from agriculture and poor sanitation practices in the watershed.

Coincident with the city's application for World Heritage Site status, a land-use investigation produced a series of alternate scenarios for regulating future development. Because monitoring had shown that most of the nutrient inputs to the lake arrived in the inflowing streams, a plan proposed by the Harvard Design School called for linking one of these streams through an artificial wetland that already existed at the Flower Nursery Gardens on the shore of the lake (see case studies in Chapter 3).

The Jinsha River, an extremely polluted watercourse situated beside the flower nursery, dumps its untreated nutrient load directly into West Lake. The Harvard Design School plan called for a series of functional treatment wetlands to be created along the riparian corridor in the headwaters of the river, each one being located immediately downstream from either a high nutrient–use tea plantation or a raw sewage–disposing town. Most of the river's flow would then be diverted through the scenic wetland on the nursery for purification before joining another nearby stream. At the intersection of these two streams, in an area now occupied by a group of abandoned aquaculture ponds, another wetland would be constructed for final polishing before the water's discharge into West Lake.

As part of an overall watershed management plan, the enhancement of the scenic wetland into a combined water treatment and tourism site achieves two goals. It is in keeping with local concerns about preventing further degradation of West Lake and at the same time promotes destinations other than the over-visited official ten scenic spots which have long been heralded in poetry.

REFERENCE

Steinitz, C., R. Peiser, and J. Xia. *Nature and Humanity in Harmony: Alternative Futures for the West Lake, Hangzhou, China.* Cambridge, MA: Harvard University Graduate School of Design, 2001.

Chapter Three

SITE-SPECIFIC LANDSCAPE ARCHITECTURE

In the past, most wetland creation projects were instigated solely for utilitarian purposes as innovative solutions to engineering problems. The potential for ancillary benefits was rarely addressed. Today, the discipline of wetland creation is on the cusp of a major shift, one in which landscape architects are performing a leading role. Recognizing and integrating elements that capitalize on the multi-faceted roles that wetlands play, both ecologically and culturally, is a challenge for landscape architects, but one that holds many rewards. Applying the principles outlined in this section will help toward achieving that end.

Design Guidelines

As we saw in the previous chapter, created wetlands can be resilient systems effective at removing contaminants and providing benefits to people and wildlife. To be truly effective, however, wetlands must be carefully designed, constructed, monitored, and maintained. This section on the landscape architecture of wetlands pertains to the creation of surface flow wetlands, and does not deal with sub-surface flow wetlands.

Creation

EXPECTED REMOVAL PERFORMANCES

The following performance summary was drawn from collating 226 input-output differences in chemical concentration from studies on created wetlands. (BOD = Biological Oxygen Demand.)

Removal Performance Variability

On average, three-quarters of all analyses indicate that contaminant removal performances of 75 percent or greater are possible. However, considerable variability exists among different chemicals in the efficacy of their removal by wetlands (3-1).

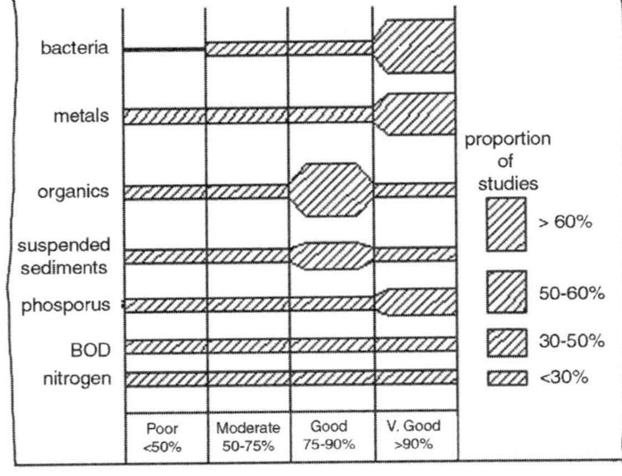

3-1

Input Loading Rate Affects Removal Performance

The ability of wetlands to continue to effectively perform as chemical traps is dependent upon the concentration and loading rate of chemicals (3-2).

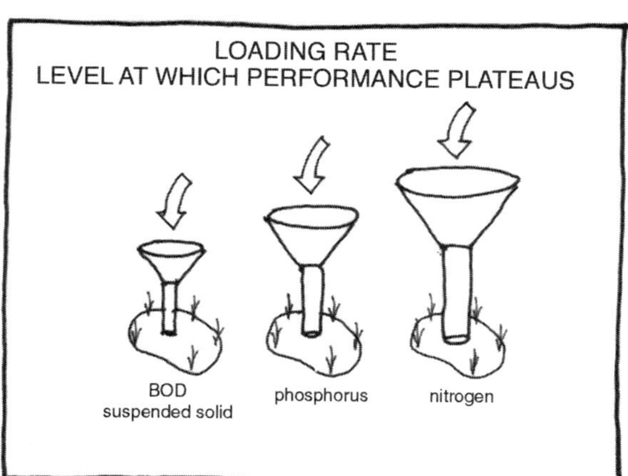

3-2

TIME-DISTANCE MODELS

The following design guidelines are drawn from fifty-two surveys of changes in the temporal and spatial concentrations of chemicals during transit through treatment wetlands.

Transport Time Effectiveness

To insure effective removal of most contaminants, water should take 10–15 days to flow through treatment wetlands (3-3).

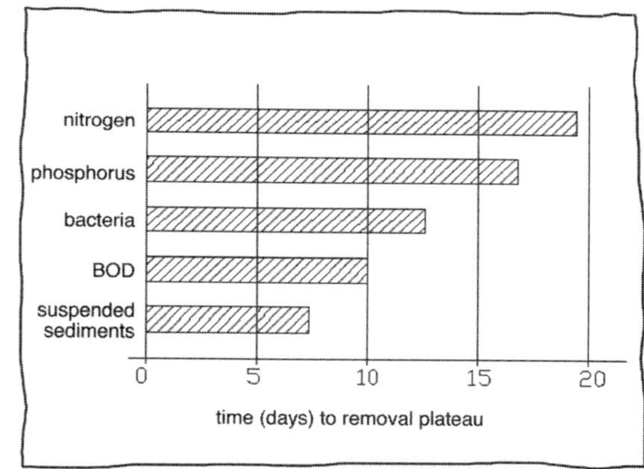

3-3

Number of Treatment Cells

Most contaminants will be effectively removed after passing through the first three treatment cells in an artificially created wetland with subdivided compartments (3-4).

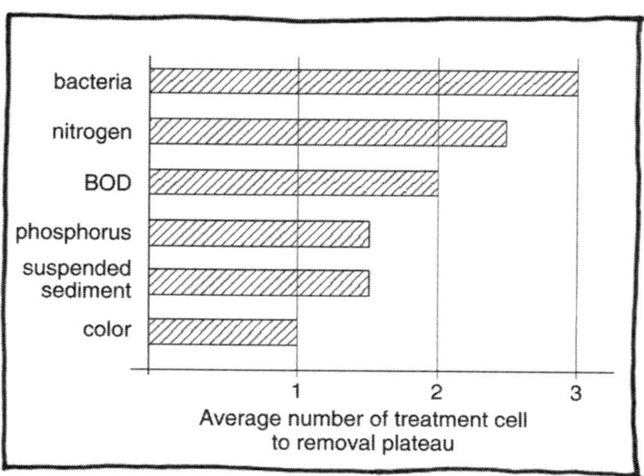

3-4

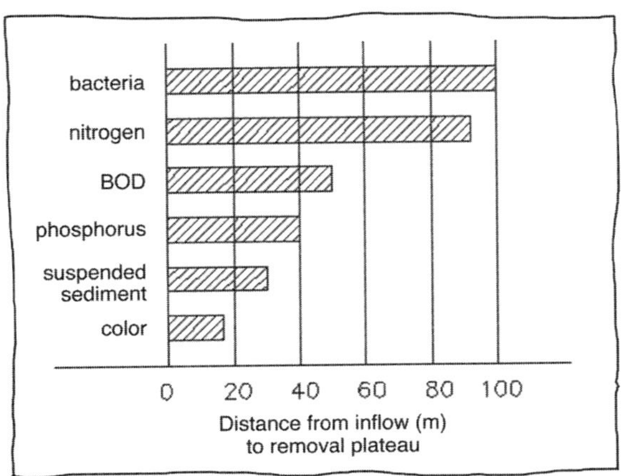

3-5

Transport Distance Effectiveness

Treatment distances from 60–120 ft. (20–40 m.) are sufficient for the effective removal of most contaminants during passage through a wetland. However, for the removal of some aqueous chemicals and bacteria, treatment distances of up to 300 ft. (100 m.) may be required (3-5).

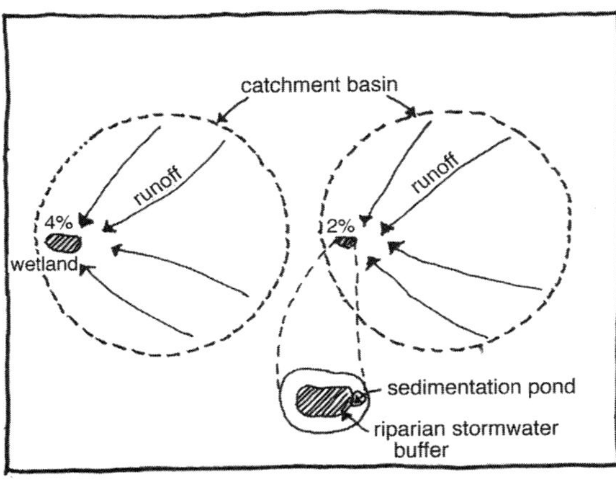

3-6

SIZE

The efficacy of created wetlands generally increases with size.

Stormwater Runoff

Wetlands should be sized at 2–4 percent of the contributing watershed area. This can be reduced to 1 or 2 percent if the design incorporates a pretreatment sedimentation pond before the wetland or provides for extended detention benches in the surrounding riparian buffer zone (3-6).

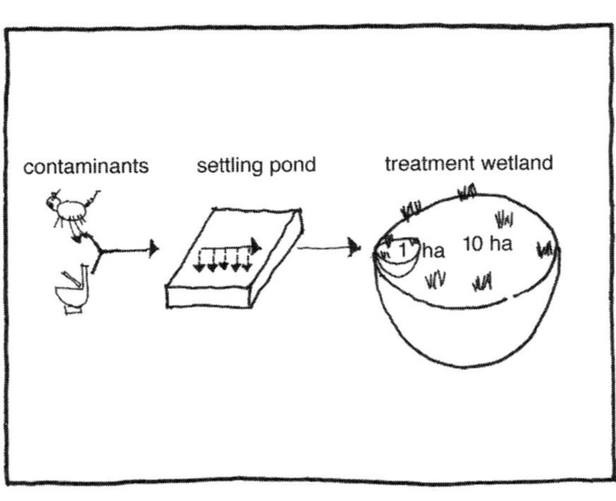

3-7

Contaminant Removal

To estimate the precise size of a created wetland, information is needed on contaminants in terms of the amounts produced, the amounts removed in pretreatment, and the estimated loading rates to the intended wetland. Most treatment wetlands that have been constructed range in size from 2.5–25 acres (1–10 ha.) (3-7).

Domestic Wastewater

To effectively treat kitchen gray-water and human sewage, a three-bedroom house will require about 900 sq. ft. (100 sq. m.) of treatment wetland, a footprint about the size of a garage or half the house.

Wildlife Attraction

Restoration projects based on reestablishing centers of biodiversity will have the greatest success if the wetlands are 1.3–10 acres (0.5–4.0 ha.) but not less than 0.05 acres (0.02 ha.) in size (3-8).

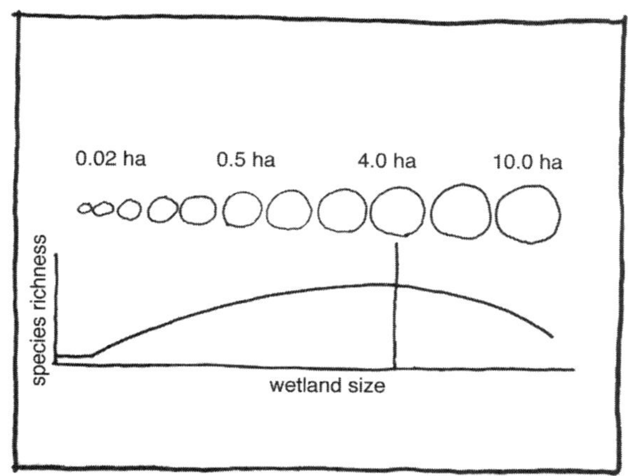

3-8

More for Less

For sites constrained by space, the water volume of the wetland area can be increased by increasing the extent of surface edge contact through design of complex internal microtopography (3-9).

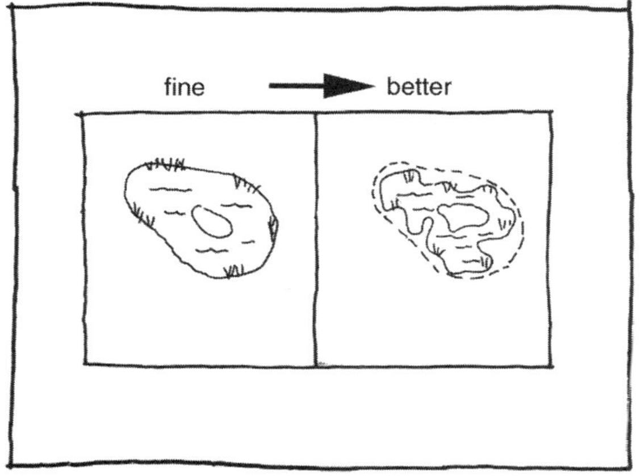

3-9

One Large or Many Small?

Water quality and quantity functioning are cumulative so that even very small wetlands will provide benefits to the overall system. Several small wetlands offer greater opportunities to avoid sensitive areas than does a single large wetland (upper panel). Often, expansion of the number of small wetlands will provide a greater benefit to wildlife than enlargement of larger wetlands because it is the land-water interface that is most important (lower panel) (3-10).

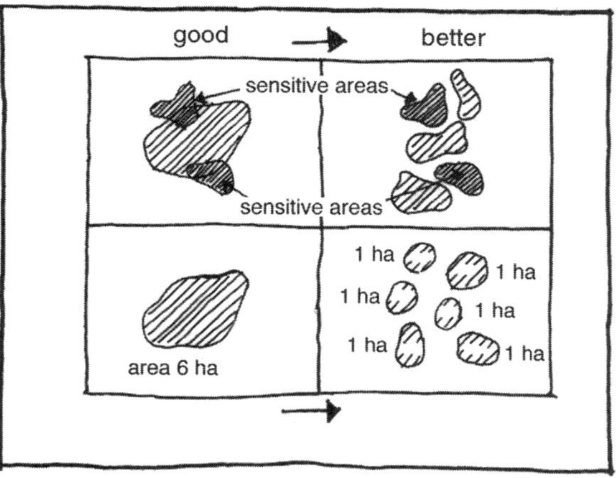

3-10

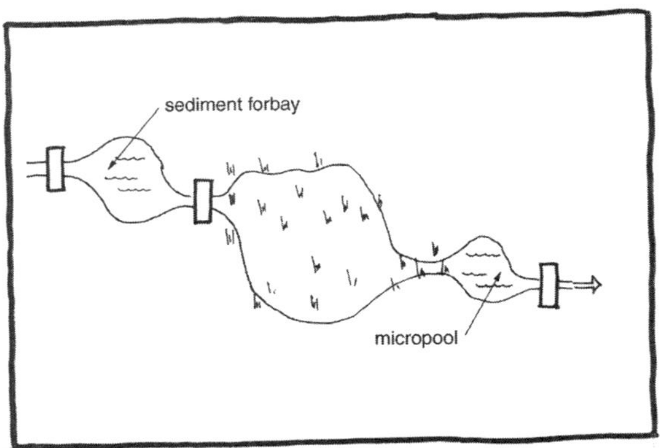

3-11

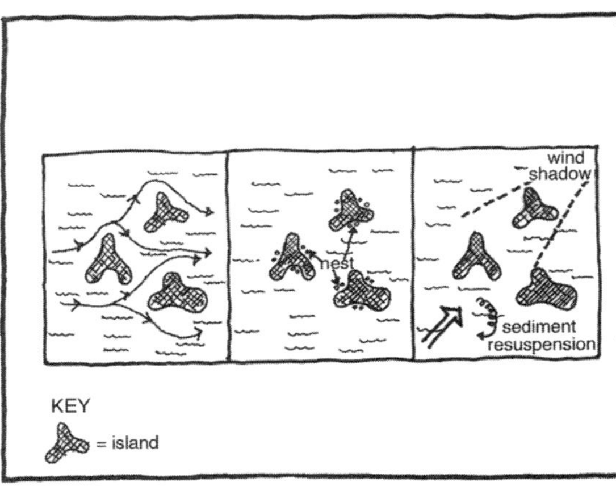

3-12

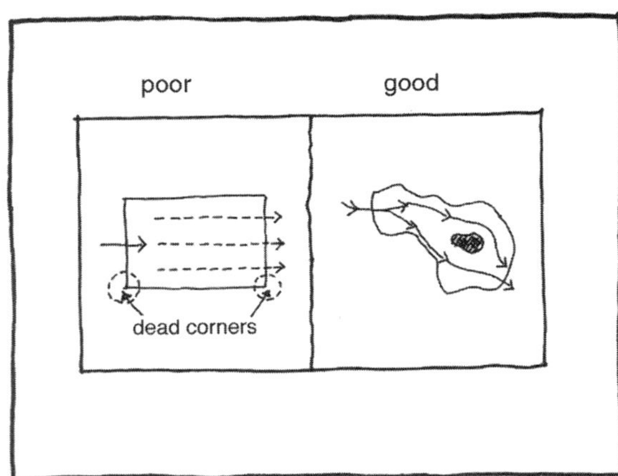

3-13

SHAPE CONFIGURATIONS

Although variations in site topography and hydrology will ultimately limit which shapes for created wetlands are realistic, within the bounds of those limitations a myriad of opportunities exist for exciting and imaginative landscape artistry that can augment wetland function.

Features

(a) Elements. A pretreament sediment forebay will serve to dissipate the energy of incoming stormwater pulses and thereby reduce particulate loading of downstream wetlands. A micropool situated at the discharge outlet will reduce the downstream release of sediments and floating organic matter (3-11).

(b) Microtopography. Multiple braided channels and islands promote water storage by increasing flow travel time which will improve effectiveness in reducing peak flows and improving water quality (left panel). Islands (more than 0.05 acres [0.02 ha.] in size) provide sanctuary for wildlife from predators and humans. Low, irregular-shaped islands are ideal as waterfowl habitats because their shape and increased edge allows distinct territories, thereby maximizing the use of space (middle panel). Islands also serve to reduce wind fetch and thereby decrease the opportunity for resuspension and downstream transport of contaminated sediments (right panel) (3-12).

Geometry

(a) Edges. It is preferable to mimic natural systems and avoid over-engineered designs with rectangular basins, hard edges, rigid channels, and regular morphology. In addition to aesthetic considerations, using curved shapes eliminates right-angled corners, which tend to be "dead water" areas in terms of contaminant removal (3-13).

(b) Width. Some studies have suggested that the widest portion of the wetland should be at the inlet end so as to facilitate a uniform and even distribution of flow. In such wetlands, designed to retain water, sediments, and/or contaminants, a constricted outlet is required in the form of a tapered end that is less than one-third the average width of the rest of wetland. In contrast, for wetlands designed to function as food sources for downstream wildlife nurseries and spawning grounds, a wide outlet is essential to promote high flushing (3-14).

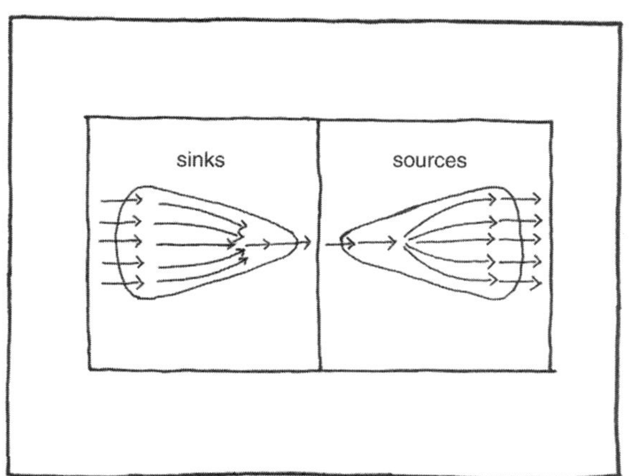

3-14

Orientation

In cold climates, maximizing southern shoreline exposure through an east-west alignment can be beneficial for wintering waterfowl using emergent vegetation as shelter (3-15).

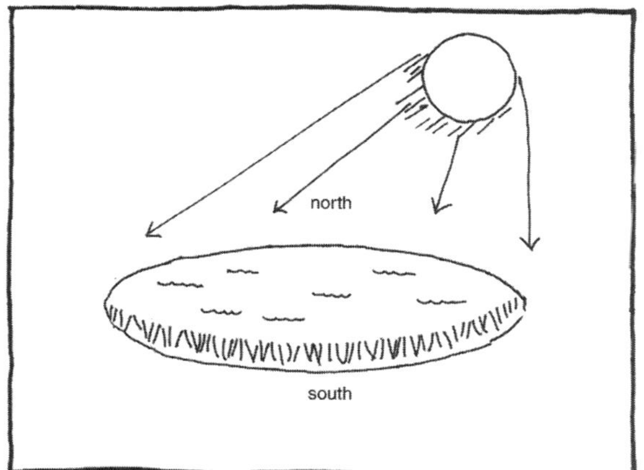

3-15

Shoreline

Creating irregular shorelines can increase their length by up to 10–20 percent per unit of wetland area. Maximizing shoreline length will provide benefits to wildlife by increasing nesting and resting areas and for water quality by increasing contact sites for contaminant removal (3-16).

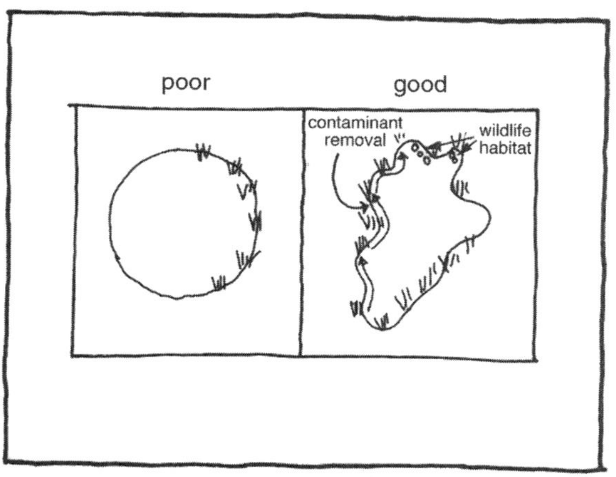

3-16

Riparian Zone

(a) ***Near Shore.*** If the possibility of shoreline erosion exists, upland runoff should be directed through swales to a limited number of discharge points. In other situations, designs should allow stormwater runoff to be introduced as sheet flow over a gentle slope of grass to limit or avoid the need for a sediment forebay. The advantage of this strategy, however, must be weighed against the strategy of establishing dense shoreline plantings of shrubs and vegetation in order to discourage nuisance waterfowl. Riparian vegetation is also essential to providing shade and aquatic thermal regulation, buffering fragile wildlife wetlands from non-point source pollution, and supplying bird habitat (3-17).

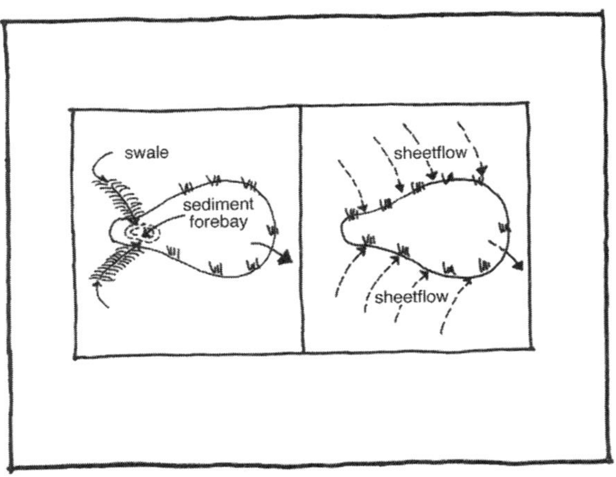

3-17

(b) ***Upslope.*** If space permits, vegetated riparian zones should be about 300 ft. (100 m.) wide to ensure adequate wildlife habitat as well as to enable animal movement between wetlands. Safety buffer zones should be constructed for extended stormwater retention and to enable groundwater recharge in the event of storms extreme enough to overflow the constructed wetland. Such temporary spring floodplains often also encourage a higher productivity of aquatic vegetation. If possible, ensure that wetland riparian forests can be connected to larger upland forests via corridors (3-18).

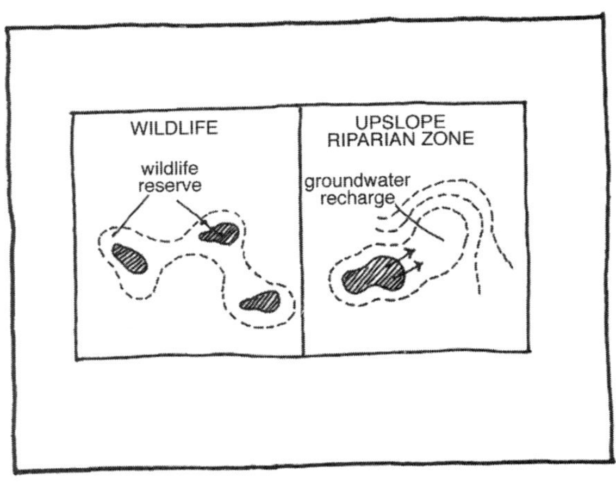

3-18

Depth

(a) Proportion. For stormwater retention, effective allocation of the wetland's total area should be on the order of 50 percent shallow (0–1 ft. [0–0.3m.]) marsh, 30 percent deep (1–3 ft. [0.3– 1.0 m.]) marsh, and 20 percent deep (3–6 ft. [1–2 m.]) water, including sediment forebay, polishing micropool, and any open water pools. For contaminant treatment wetlands, it is recommended that very shallow (less than 6 in. [15 cm.]) depths comprise up to 50–70 percent of the total surface area in order to promote even sheetflow. Wetlands created as wildlife preserves should have a near equal ratio of shallow (less than 0.5 ft. [0.2 m.]) vegetated and mudflat areas to deeper (1–3 ft. [0.3–1 m.]) open-water areas. In cold climates, all depths should be increased to permit water flow under the ice during the winter (3-19).

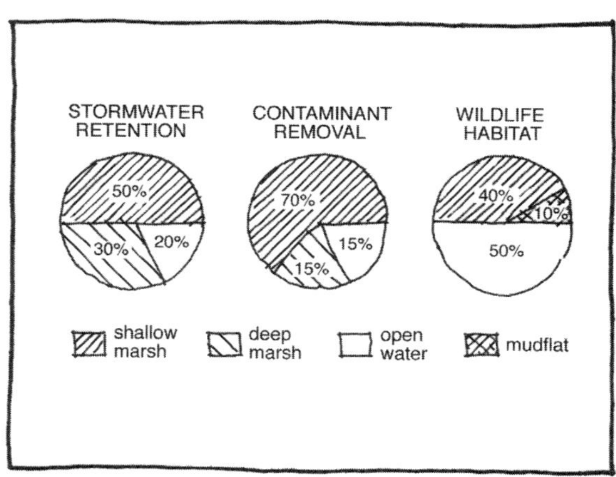

3-19

(b) Arrangement. Deep zones should be arranged perpendicular to flow direction to limit the chance that water can short-circuit the intended pathway. Because a mixture of flooding regimes expands wetland functions, interspersion of shallow and deep water areas will increase the efficiency of contaminant removal and attractiveness for wildlife. For example, wetlands created for the downstream export of food in the form of organic matter require that some regions dry out and decompose rapidly before the next spring flooding (3-20).

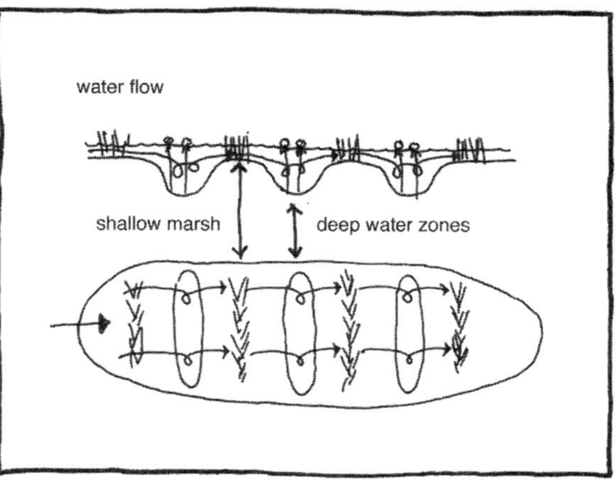

3-20

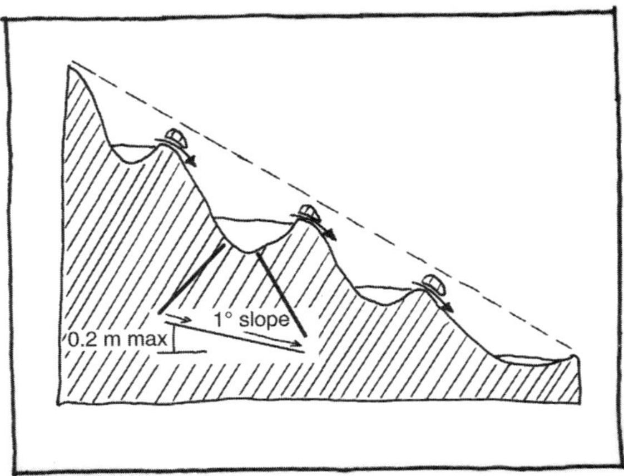

3-21

Slope

(a) Longitudinal. Slopes should not exceed 0.5–1.0 percent in order to maintain depth differences of less than 0.5 ft. (0.2 m.) from inlet to outlet. On level sites, treatment cells can be created with dikes and berms; on sloping sites, cells can be terraced into the landscape. For steep sites, lengthwise division of the wetland into a series of cells is recommended to prevent water accumulating at the downstream end to depths undesirable for plant viability (3-21).

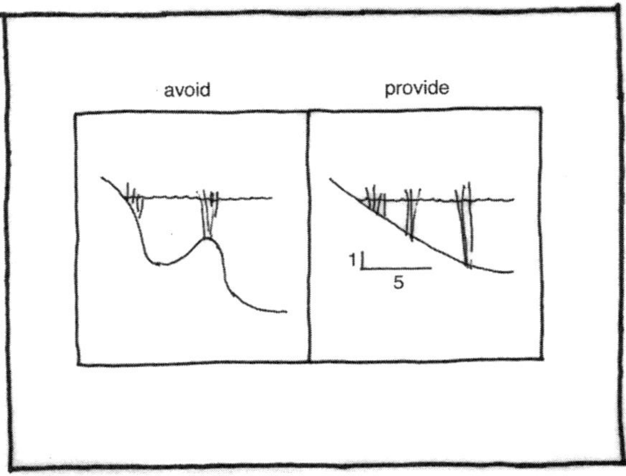

3-22

(b) Vertical. Shorelines should be shallowly graded from 3:1–5:1 to provide easy access for wildlife, to limit erosion caused by water contact with the toe of the bank, and to provide opportunity for periodic drawdown dewatering, as few wetland plants require permanent flooding (right panel); deep holes or steep drop-offs should be avoided to minimize hazards to children (left panel) (3-22). Alternatively, presence of a marginal trench will protect openwater areas from being colonized by spreading shoreline plants (3-23).

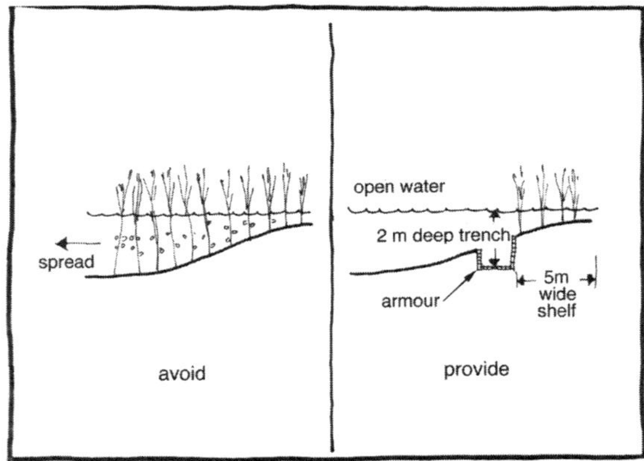

3-23

Internal Partitioning

(a) Performance. Because complex internal edges of wetlands will increase the overall planted surface area, the longest possible flow path should be created so as to maximize contact of water with plants. For medium-size wetlands, use of intermediate path distributors will help to reestablish dispersed flow patterns through the dense vegetation beds. For large wetlands, use of parallel cells reduces the opportunity for the development of preferential flow paths that could lead to flow short-circuiting and decreased contact of contaminated water with plant surfaces (3-24).

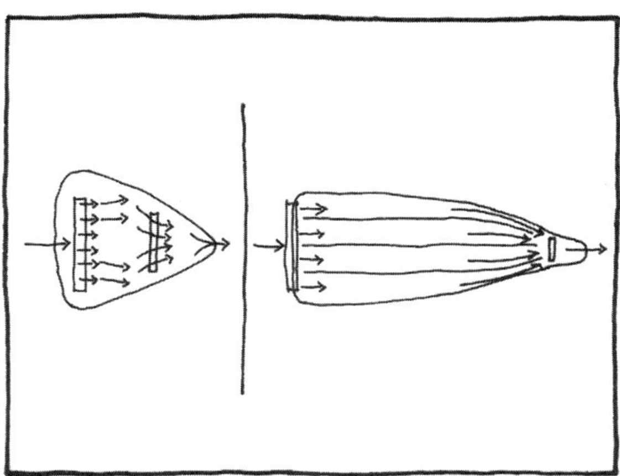

3-24

(b) Operation. Multiple cells also facilitate easy operation and maintenance because some cells can be temporarily shut down, cleaned out, or replanted as necessary without disrupting the entire treatment system. An additional benefit is the isolation of particular cells to limit spread of problems such as plant disease or extreme toxic spills (3-25).

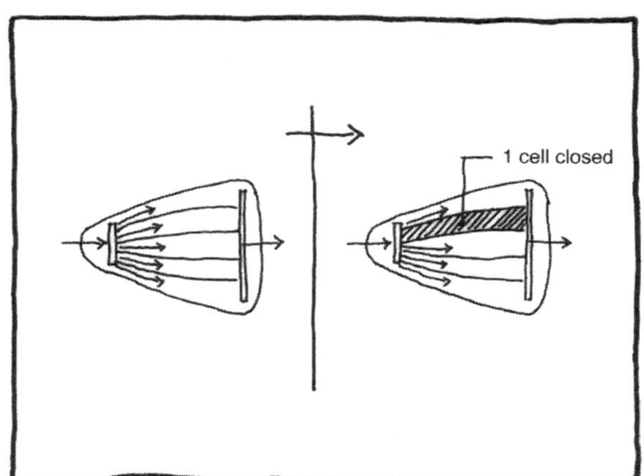

3-25

Length:width Ratio

In theory, length:width ratios of greater than 2:1 (and preferably 3:1–4:1) appear most efficient for contaminant removal. In reality, an optimal length:width ratio of about 2.5:1 represents the balance between removal performance and increasing costs associated with berm construction (3-26).

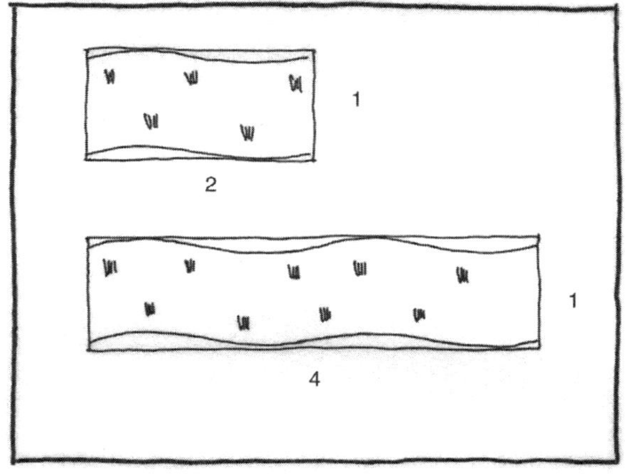

3-26

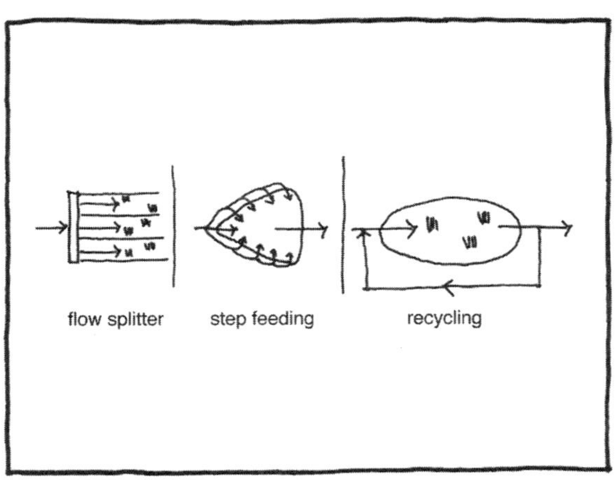

3-27

Water Regulation

Treatment performance is optimal when flow is evenly distributed from the beginning to end of the cell. In general, the smaller the length:width ratio of the wetland, the more important equalized inlet flow distribution becomes. Flow splitters can be used for wetlands comprised of parallel cells. The use of multiple input points along the length of cells (termed "step-feeding") is a strategy to distribute contaminant load evenly, limiting toxicity at the upper end. Water recycling increases detention time and therefore contaminant removal, in addition to maintaining adequate flows during droughts and compensating for evapotransporation losses (3-27).

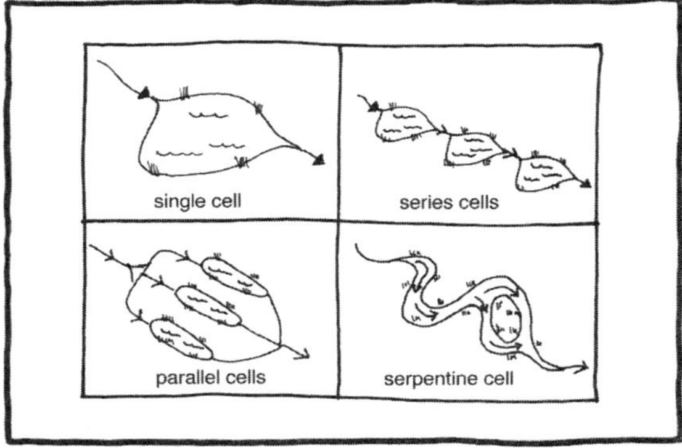

3-28

Spatial Arrangement

The interlinked networking of cells within an individual wetland or the specific associations of grouped wetland features within a larger combined water-wetland system will strongly influence resulting performance.

Alternate Configurations

Achievement of project goals through increased wetland complexity may often have to be balnced in relation to space constraints (3-28).

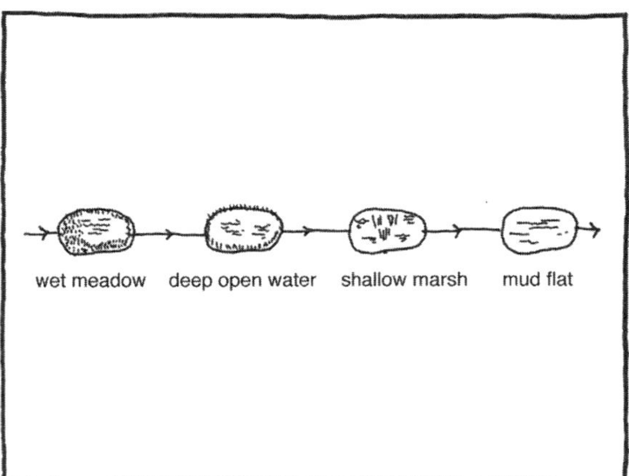

3-29

Sequencing

Changing water depth along a series of created wetlands will promote establishment of a suite of different plants and thereby increase the diversity of both wildlife habitats and contaminants that are removed (3-29).

Multiple-stage Treatments

(a) Series-combinations. Creation of a comprehensive water-wetland treatment system, alternating wetlands with openwater ponds and subsurface filters, will increase the likelihood of project success. For regions where groundwater recharge is of importance, the final wetland cell can be left unlined to promote infiltration of treated water. In northern regions where treatment performance is believed to decrease at low temperatures, it may be necessary to store the winter load in a lagoon for discharge into the wetland during the open-water season (3-30).

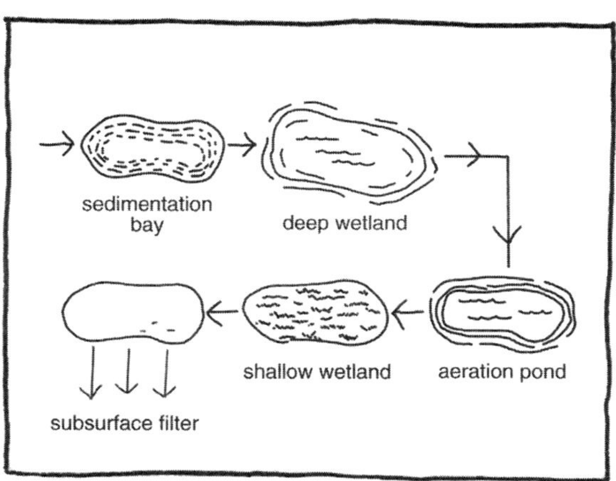

3-30

(b) Use of the surrounding landscape. Additional incorporation of natural offsite features such as upslope filtering swales or final polishing wetlands into a comprehensive functional design plan will provide further benefits (3-31).

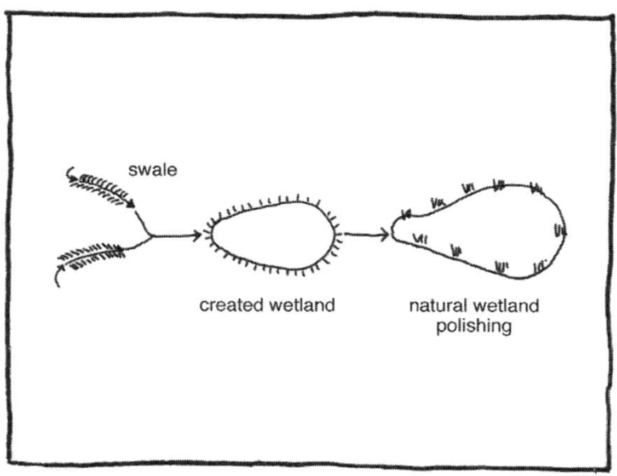

3-31

Multiple Objectives

Never attempt to incorporate all desired wetland uses into a single encompassing water-wetland system. For example, to limit the possibility of contaminant transfer to wildlife, ensure that the surface area of serially-linked treatment systems progressively increases. This will attract animals to the larger, downstream wetlands where water quality will be higher (3-32).

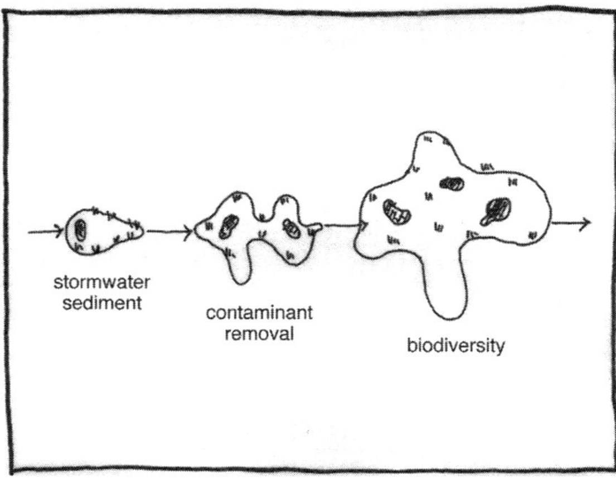

3-32

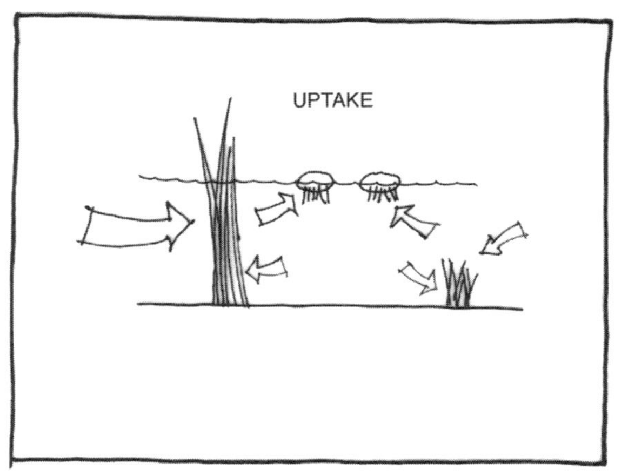

3-33

SUITABILITY OF AQUATIC VEGETATION

Intended Role

Wetlands are created for a variety of reasons, the choice of which will influence the initial screening and selection of suitably appropriate vegetation.

(a) Chemical sequestration. Plant species vary in their effectiveness in binding different contaminants. For example, chemical uptake is often highest in emergent species such as cattails and bulrush (3-33).

3-34

(b) Habitat value. The role of vegetation as a food and cover resource for birds, mammals and other aquatic animals is species-specific (3-34).

3-35

(c) Attractiveness. Aesthetic quality of created wetlands depends on careful plant selection (3-35).

(d) Flow attenuation. Stands of different plant species will restrict the movement of water and wind variably. This is important for concerns of stormwater retention and contaminated sediment resuspension (3-36).

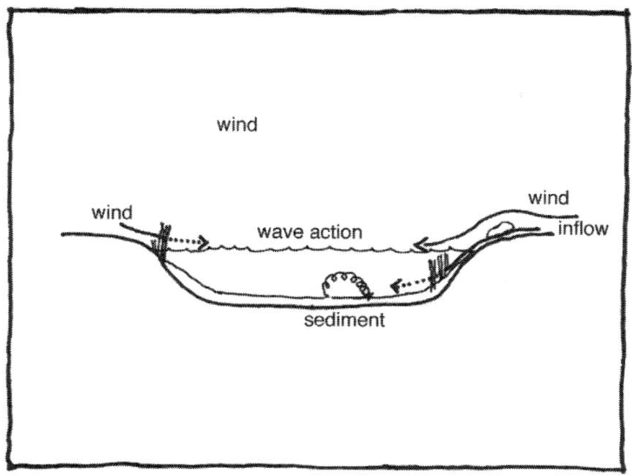

3-36

Vegetation Options

Given determination of the intended role for the created wetland, the next decision entails consideration of what broad type of plant material should be selected for use.

(a) Planting zones. Six major planting zones exist in any wetland habitat in relation to normal water level (3-37).

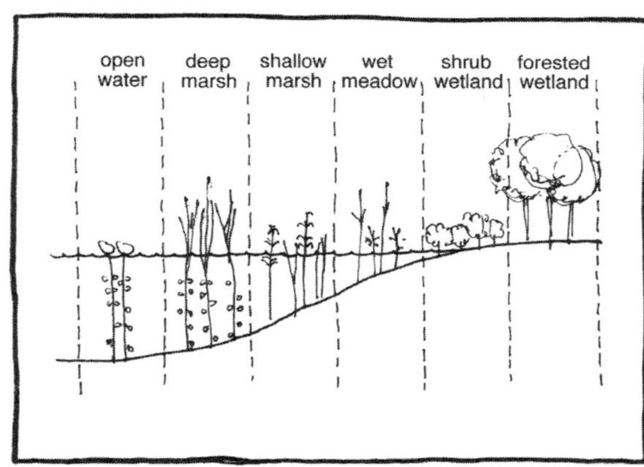

3-37

(b) Growth forms. Five major growth forms of vegetation are available for use in created wetlands in relation to their suitability for particular planting zones (3-38).

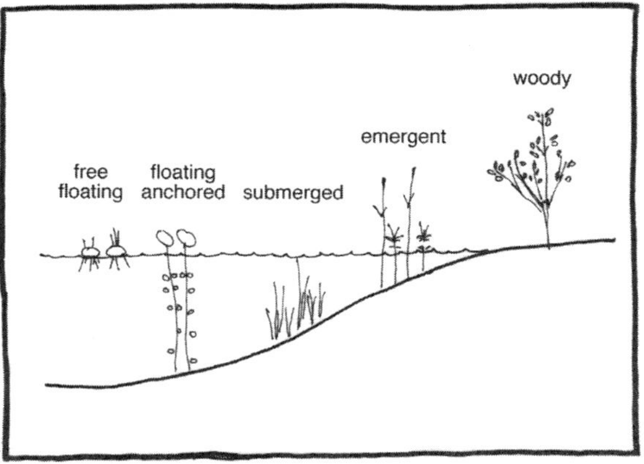

3-38

3-39

*(c) **Growth efficiencies.*** It is important to take into consideration the projected growth efficiencies of the different types of vegetation under review. Competitive interactions, invasiveness, productivity, and biomass accrual will all affect the resulting function and appearance of the created wetland (3-39).

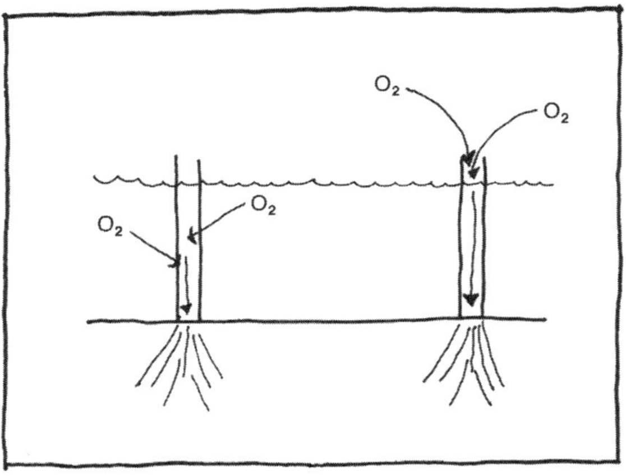

3-40

Environmental Determinants

Particulars of individual species selection must be made in association with an understanding of the anticipated environmental conditions of the created wetland. Such precise information is dependent on the services of an experienced wetland horticulturist.

(a) Sediments. In water-saturated soils, oxygen needed for root respiration must be obtained from the aerial parts of emergent plants either by passive diffusion or by convection. Increasing the organic matter content of bottom sediments helps to bind contaminants (3-40).

3-41

(b) Water tolerance. Species of emergent herbaceous and woody plants have widely different tolerances to water inundation (3-41).

(c) Seasonality. Plant species vary greatly in terms of their temperature- and light-regulated periods of growth and dormancy. In most regions of North America, growth is likely for only six to nine months of the year, and in consequence, wetland performance may vary seasonally.

(d) Longevity. The long-term survival of plants will be influenced by the aqueous chemistry of the wetland with respect to both growth nutrients and contaminants. Especially important when choosing plants for stormwater wetlands is the issue of their tolerance for salt, due to its widespread use as a deicing agent.

(e) Soil characteristics. Planting success is dependent on soil type and chemistry.

(f) Hardiness. Plant hardiness—ease of germination and propagation—will ultimately determine the resulting floral composition of the wetland. Disease resistance will be maximized through selection of a diverse planting palette (3-42).

3-42

(g) Hydrology. The exposure to wave energy and the projected flow rates and patterns of water circulation through the created wetland will affect which types of plants will thrive or fail (3-43).

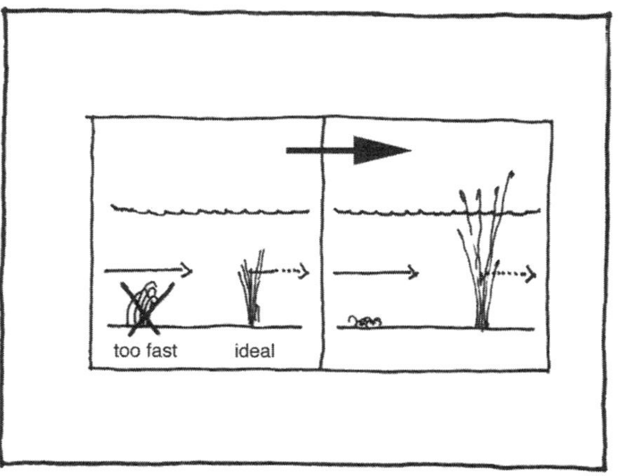

3-43

Design Strategies

The spatial arrangement of planted vegetation can play a dramatic role in the functional efficacy as well as aesthetic quality of created wetlands.

(a) Spacing. Planting success may be dependent on the spacing interval recommended for particular species.

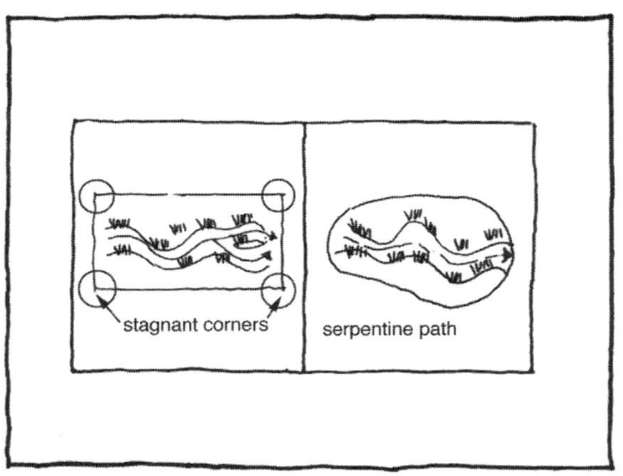

3-44

(b) Water routing. For wetlands intended as chemical sinks, the planting configuration will determine water routing and therefore the opportunity for vegetation to bind contaminants. Blind corners or "dead water" areas should always be avoided, and high marsh plants can be used during dry weather to redirect flows (3-44).

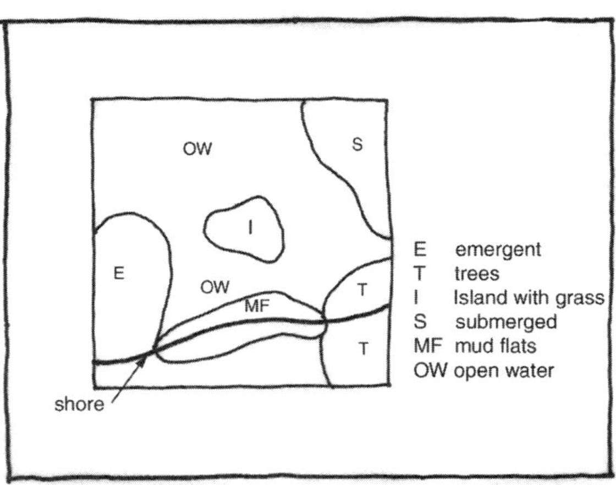

3-45

(c) Habitat. For wetlands intended primarily as centers for sustaining wildlife, the optimal planting configuration should be one that maximizes habitat diversity (both vertical and horizontal) as much as possible. This can be accomplished through both the spatial arrangement of broadly different vegetation compartments within and around the wetland, and by establishing localized site-specific centers of high structural complexity. In many cases, a mosaic of small patches of emergent vegetation interspersed with openwater pools, channels, and mudflats rather than the creation of several large centers of different habitat will maximize habitat diversity. Planting beds at least 3 ft. (0.3 m.) high and several yards (meters) wide is important for providing spring nesting cover (3-45).

(d) Beauty. For wetlands designed as visual spaces, color, bloom season, balance, and spatial arrangement of different plants are all important attributes to consider, as is the wetland's look in future years of natural succession. Successful water gardens employ artistic imagination in establishing the relationship between vegetative and aqueous landforms (3-46).

3-46

AN IDEAL PLANT?

A narrow planting list may sometimes be beneficial, with all selected for the following optimal traits: wildlife value, contaminant uptake ability, broad hydrological tolerance, rapid establishment, and bank stabilization. Invasive plants or those that are preferentially foraged by waterfowl should be avoided.

Project Development

PLANNING

Careful attention must be given to the following framework of planning criteria to ensure that a wetland construction proposal will receive official approval and achieve its intended success.

Site Selection

Topics to consider during initial screening of potential construction sites include proximity and elevation in relation to source of contaminated water, prior environmental disturbance, distance from existing water bodies and human habitations, location of floodplain and depth of water table, number and location of outlets, low elevation gradients for promotion of sediment and contaminant deposition, land-use patterns, ease of access, presence of threatened or endangered species and historical sites, perceived ability to capitalize on existing landscape features so as to limit earthmoving, nature of underlying substrate, and microclimatic variables such as estimated solar and wind exposure.

Identification of Assets and Requirements

For the selected site, it is important to have a priori understanding of such issues as the amount and type of land available, soil and water resources, equipment, financial backing, and extent of expertise before any project is initiated. In terms of hydrology, for example, it is essential to investigate whether the water supply is year-round or only seasonal, whether pumps and pipelines may be required, and what legalities might be involved in obtaining the water.

Refinement of Project Goals

Site particulars pertaining to water supply and local topography will strongly dictate the type of ensuing wetland that can be constructed. For example, issues for wetlands created on uplands, slopes, or lowlands include water supply, erosion potential, drought susceptibility, groundwater permeability and pollution, wildlife habitat, and connectivity to other wetlands.

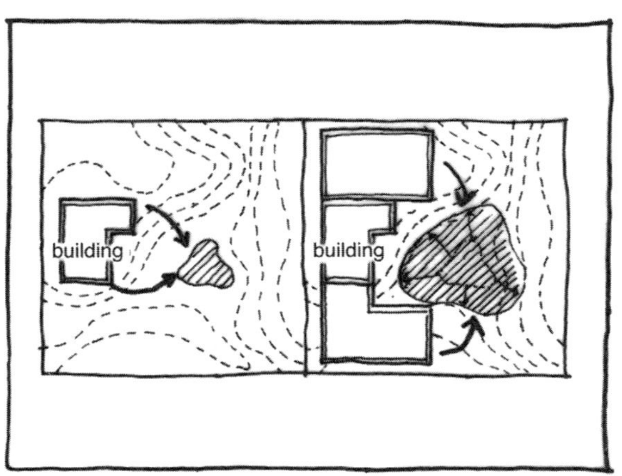

3-47

Site Inventory and Plan Formulation

Site maps with details on surface topography, water table depth, soil type, and existing vegetation patterns provide a foundation on which to overlay a series of alternative design plans. Given the luxury of space, plans should accommodate possibilities for future wetland expansion should the need arise, as, for example with increased watershed development and consequent stormwater runoff. Attention to a detailed site grading plan is essential (3-47).

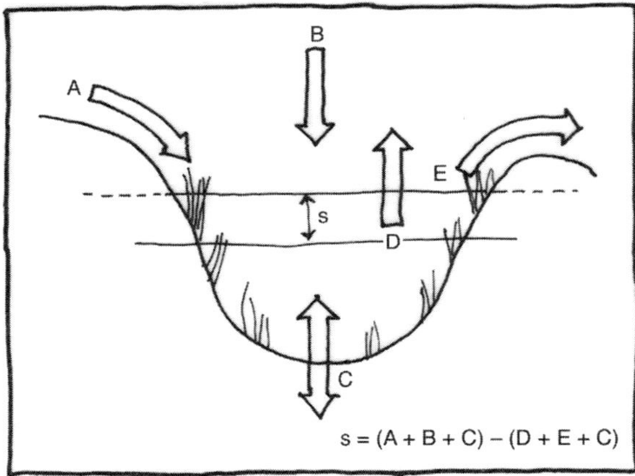

3-48

Hydrologic Considerations

(a) Water balance model. A water budget(s) should be calculated relating net storage capacity to the sum of surface inflows (including contaminated inputs), rainfall contributions, and groundwater exchanges, minus the losses from outflow and evapotransporation. With this information, it is possible to manipulate the water budget should the need arise. For example, increased storage capacity can be achieved by varying outflow rates, deepening pools, and reducing evapotransporation through riparian tree shading, wind obstruction, and plant alteration (3-48).

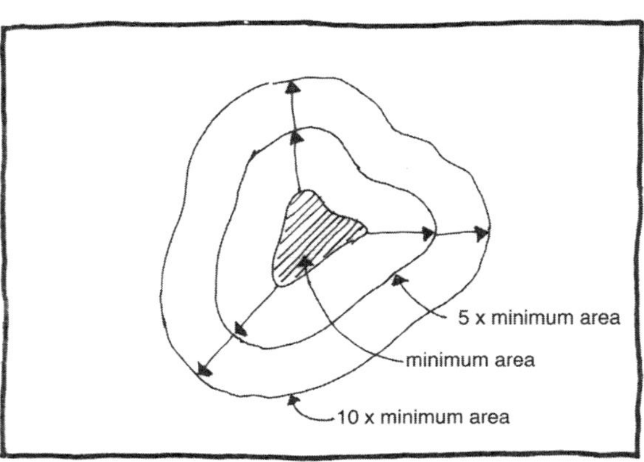

3-49

(b) Quick estimation. The ultimate goal is to calculate the minimum water levels needed in addition to identifying the structural outlet devices required. Where such detailed budget models are unpredictable, the wetland should be sized to accommodate 5–10 times the minimum amount of water required for sustained plant growth. Once the wetland becomes operational, excess water can be drained away using a control outlet (3-49).

Environmental Clearances

Social issues are often more difficult to address than technical or scientific issues. The labor-intensive, time-consuming, and occasionally frustrating process of obtaining justifiably cautious approval for wetland creation projects should be initiated as soon as possible.

Design Wetland

Once the water surface elevation of the planned wetland is selected, the design process begins. The design should be as synchronous as possible with the landscape, rather than imposed on it. Design features that relate to the surrounding landscape, such as an extended stormwater detention zone or a network of feeder swales to control erosion and direct overland flow, should be incorporated into the overall plan. Wetlands should be designed to operate with a minimum of maintenance and be kept simple to avoid reliance on complex technical systems. A decision must be made at the start as to how natural or artificial (not engineered) the final wetland should look (3-50).

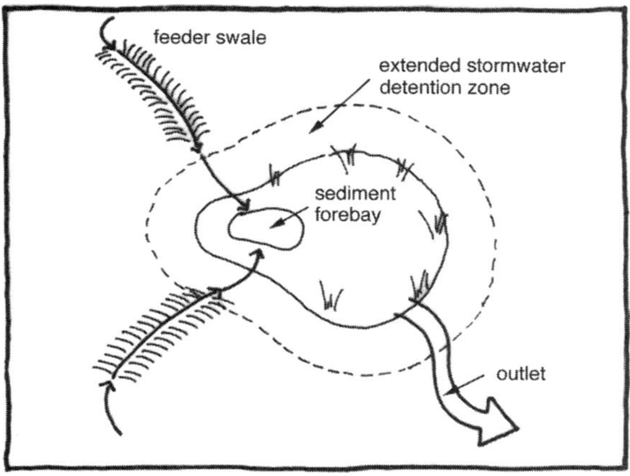

3-50

Design Levee System

Given the planned water depth, the levee system should be designed with an additional 3–6 ft. (1–2 m.) of freeboard for wave action and to accommodate extreme storm events (3-51).

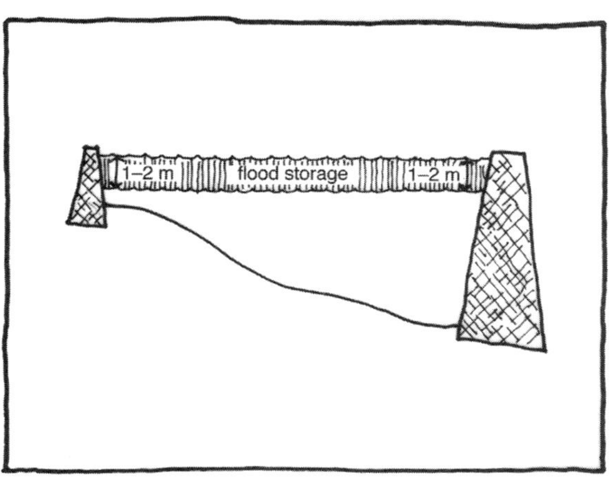

3-51

3-52

Select Water Control Structure

Outlet structures to regulate water levels are essential to a wetland's proper function, and require careful attention paid to accompanying backup safety options. Outlets should be simple, easy to adjust, and flexible enough to be changed if the need arises (3-52).

Financing

A detailed budget should be prepared that considers all particulars including survey inventory, design plans, materials and labor, modeling, monitoring and management, legal fees, and a cash buffer to cover any unforeseen expenses.

Timing

A phasing sequence needs to be developed that clearly lists all stages from survey to post-construction management.

SITE PREPARATION AND CONSTRUCTION

This description of construction sequencing pertains to design projects of any size.

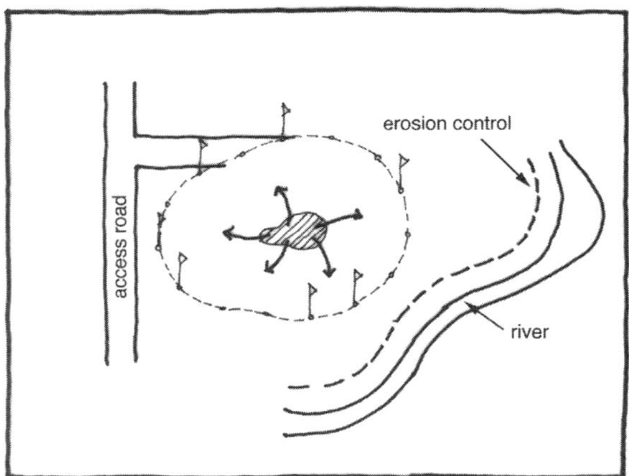

3-53

Marking, Access and Excavation

Careful surveying is required to delineate the location of the selected site in relation to access and erosion control measures during construction. Excavation should proceed from the center towards the intended land edge, saving the soil for shoreline and riparian landscaping. Banks should be reinforced with permeable material. In all cases, environmentally sensitive construction techniques must be employed, including interim water redirection if required. For some sites, it may be necessary to temporarily drain existing water during construction (3-53).

Islands

Islands should be at least 144 sq. ft. (4 sq. m.), located 45 ft. (15 m.) apart, and be separated from the shoreline by a minimum of 45 ft. (15 m.) of permanently flooded water (3-54).

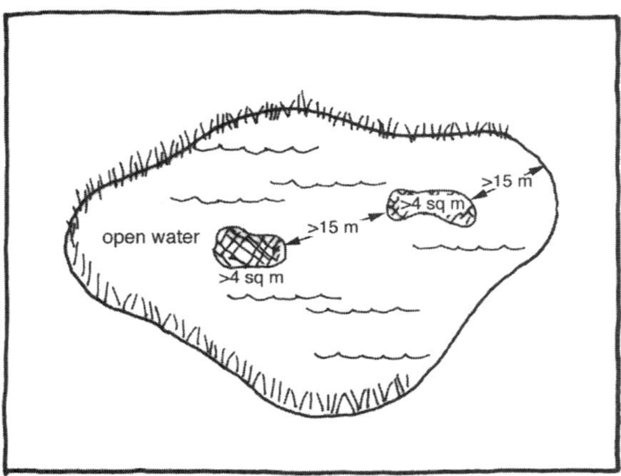

3-54

Dikes, Berms and Hydrological Control

To ensure long-term stability, sidewalls should be sloped no steeper than 2 horizontal: 1 vertical; riprapped or protected by erosion control fabric; have an adequate freeboard to accommodate occasional stormwater pulses and open-water waves; include anti-seepage collars; have inlet and outlet regulation devices; and include an emergency spillway and rocky splash areas to dissipate inflows (3-55).

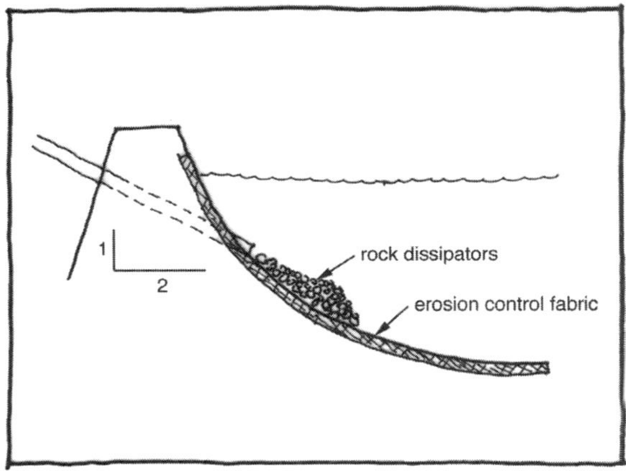

3-55

Soil Modification

It may be necessary to reconstruct the existing soil so that it is appropriate for bedding material for plants. Excavated soils can be stockpiled to use later as plant bed material.

Surface Smoothing

Following ground clearance of stones and wood, a sand layer should be laid down to further protect the overlaying impervious liner. If burrowing animals pose a problem, a plastic net layer should be added beneath the sand. At least 4–6 in. (10–15 cm.) of soil is required for shallow-rooting aquatic plants (3-56).

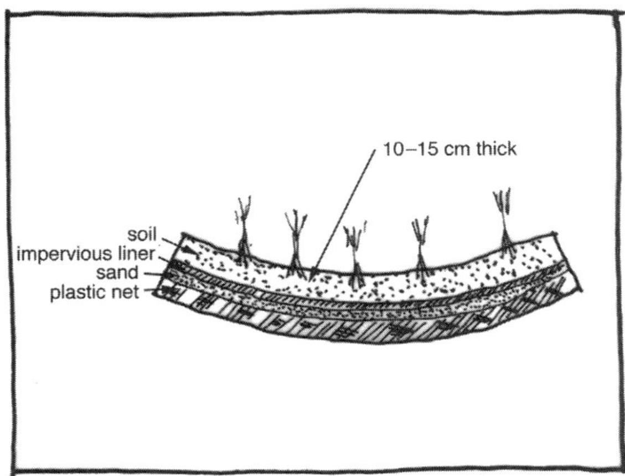

3-56

3-57

Incorporation of Physical Design Amenities
Waterfalls, water spouts, spring stones, small fountains, and recycling streams can all be worked into the construction process at this design stage (3-57).

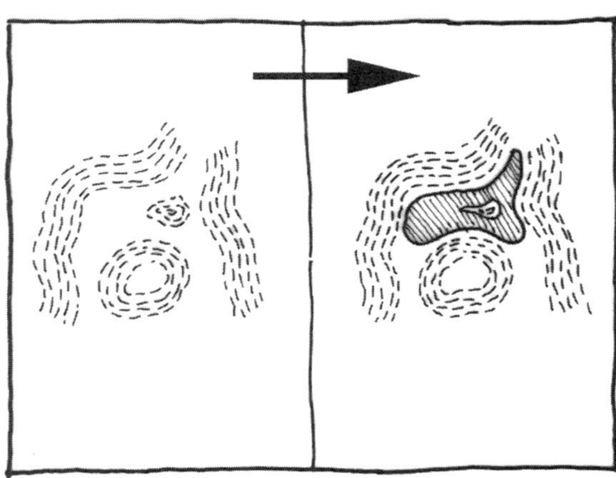
3-58

Grading
One of most important factors affecting wetland success is the creation of precise elevations. Attention should be paid to site-sensitive design where pools, basins, and channels are shaped and blended into the natural contours of the landscape (3-58).

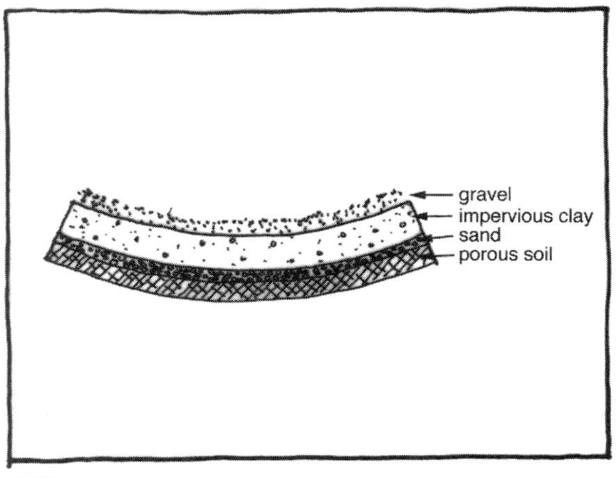

3-59

Bottom Sealing
Because surface pooling of contaminated water poses a threat to groundwater resources, care must be exercised in sealing the bottoms of constructed wetlands intended for treating contaminants. Flexible pond liners (dark colors give greater depth impression; roughed surfaces allow for settling particulates to collect for natural color) should be laid down on slopes less than 45 degrees to protect the liner from stretch. Alternatively, a 16–20 in. (40–50 cm.) layer of clay or a 8–12 in. (20–30 cm.) layer of bentonite, tamped down firmly and topped with 2–4 in. (5–10 cm.) of gravel, can be used as a bottom sealant (3-59).

Edging Borders and Riparian Landscaping
A marginal zone along the wetland edge is created using a mixture of hard and soft liner anchoring techniques such as riparian plants, gravel, natural or pavement stones, wooden platforms, bricks, or jute sacking in order to prevent shoreline erosion (3-60).

3-60

Fish Habitat
Placement of large woody debris and/or gravel spawning beds may be needed to create nursery habitats for fish (3-61).

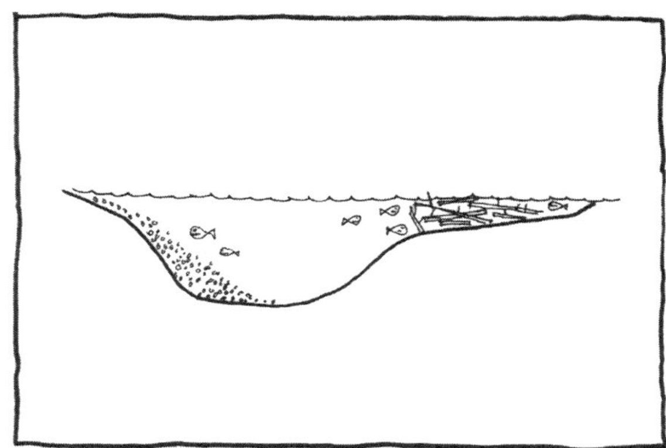
3-61

Engineering and Architectural Installations
Water features such as submerged or vacuum pumps, aeration pumps, and clean-water filters, and surface features such as bridges, stepping stones, and shoreline decks or piers are added next (3-62).

3-62

3-63

Site Preparation

Once the wetland has been shaped and graded, the compacted soil should be broken up before shallow flooding to enable the soil to settle into a level bed. In order to remove an existing seed bank and thereby reduce competition prior to spring plantings, the upper few inches of soil can be excavated (3-63).

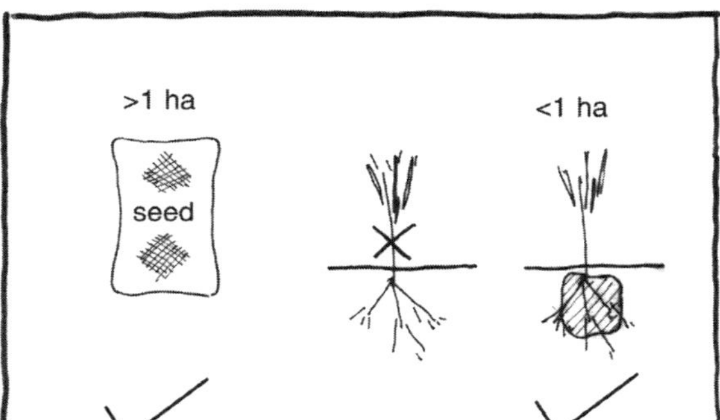

3-64

PLANTING

Type

Wetland plants come in several forms: seed banks of individual species, wetland soils harboring viable mixtures of seeds, rhizomes and tubers from many species, and entire plants suitable for transplanting. If the planting area is greater than 2.5 acres (1 ha.), it will be more economical to use seed (though the results may be less predictable). Container or peat-potted plants have higher survival rates then do bare root stock plants, so even if their initial price is higher, they are often more economical in the long run (3-64).

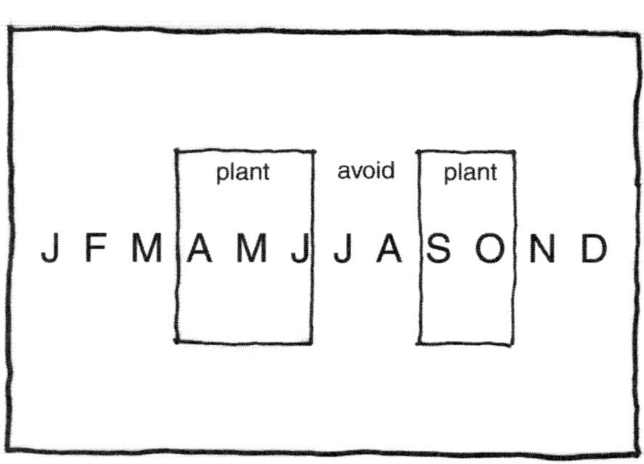

3-65

Timing

Though not essential, is is desirable to sow seeds in the spring to capitalize on a full growing season. With non-dormant spring plantings, timing of planting may be one of the most important elements in project success (so much so that if delays intervene, it is often wiser to delay planting for an entire year rather than attempt to plant one month outside the ideal window). For autumn plantings, dormant plants are advisable. They will have greater overwinter survival success and be ready to grow again in the early spring (3-65).

Spacing

A spacing interval of 3 ft. (1 m.) for emergents (about 4,000 per acre [10,000 per ha.) and 1.5 ft. (0.5 m.) for small flowering plants (about 8,000 per acre [20,000 per ha.]) will allow for spread. The planting density and stock size should be increased to compensate for losses expected from water turbulence in some high energy systems and also from wildlife (herbivores are often attracted to isolated clumps rather than uniform ground cover). Some plants are aggressive and may need to be physically constrained within pots when planted (3-66).

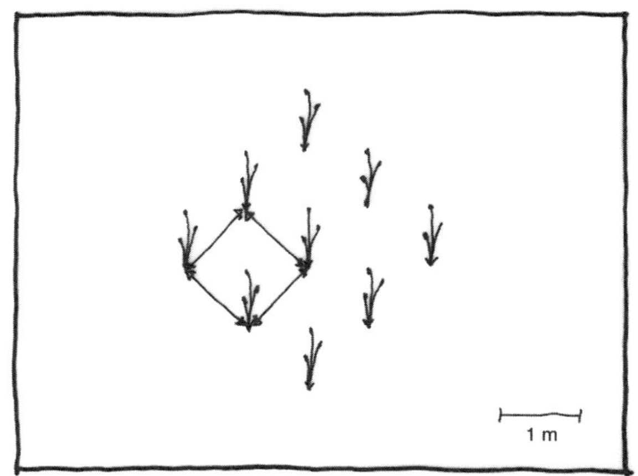

3-66

Fertilizer

In low-productivity systems, placement of a fertilizer tablet at the bottom of the planting hole will limit the chance of algal blooms developing in the water column.

Water

(a) Depth. An experienced installation crew can advise on the pluses and minuses of dry planting versus submerged planting. In most cases, 2–3 in. (5–8 cm.) deep is ideal for littoral plantings (3-67).

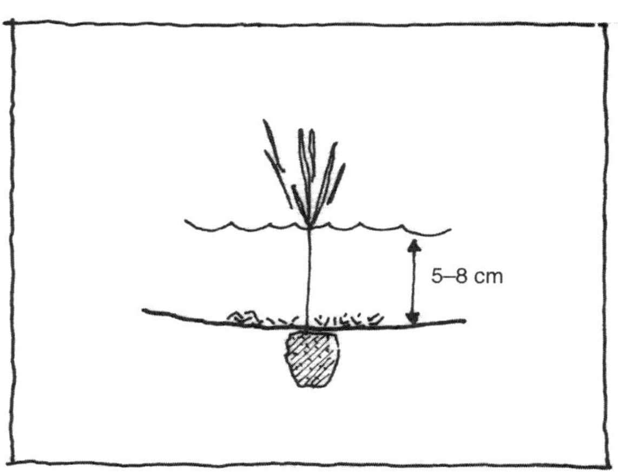

3-67

(b) Hydrology. Gradual flooding protects the liner and allows plants to become acclimatized. Plantings should be allowed to become well established before contaminated water is introduced. Some plants require water fluctuations, but limit drawdown periods until vegetation is well established (3-68).

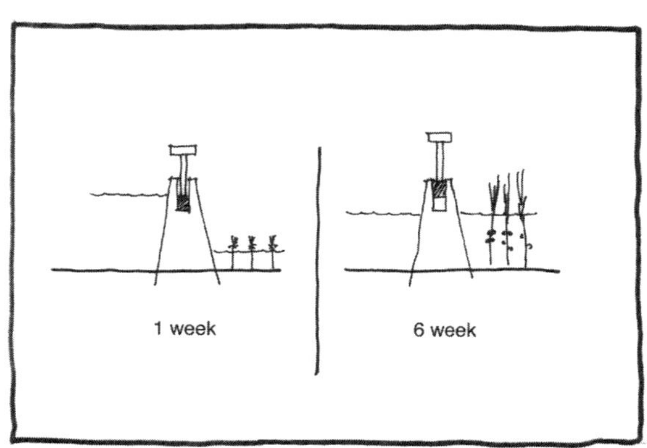

3-68

3-69

Herbivory

It is essential that screens or string fencing with stakes be built over the maturing plants to limit browsing by herbaceous animals. Because geese graze from the water towards the shore, string can be used for the outer edge of the exclosure. Along the inland edge, however, it may be necessary to use heavier materials, especially in situations where dogs are present (3-69).

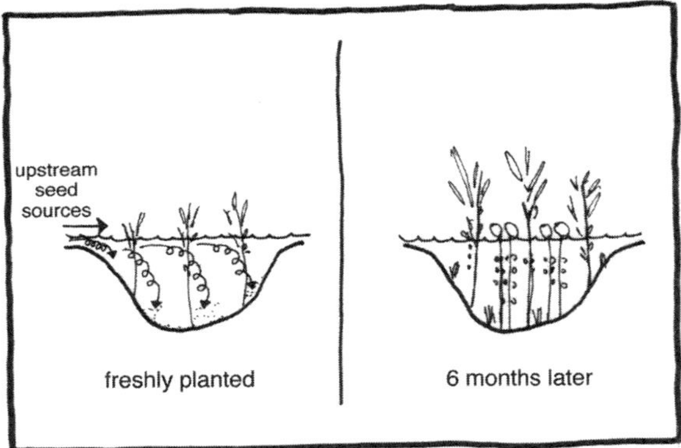

3-70

Or Natural Colonization?

In some cases, volunteer plant species have been found to contribute to over 90 percent of cover and diversity in created wetlands (depending on nearby seed sources and transport avenues, extent of existing vegetation, time of year of wetland construction, and source and type of soil). Because planting can be costly and possibly unnecessary, one strategy is to use deliberate planting only to accelerate the natural processes of revegetation (3-70).

MONITORING

Wetland creation does not end when the excavation site is filled with water over which plants are seed-sprinkled or hand-planted.

Final Maintenance Plan

Essential follow-up duties include long-term responsibilities and funding, routine cleaning and checking of inlet and outlet water control installations, mowing, embankment inspection, monitoring depth of sediment accumulation, water level and hydrological budget recalculation, water quality monitoring, tracking plant species composition, density, and viability, and checking overall wetland health.

Plant Establishment Period

Surveys for at least one full growing season are essential and should be part of any landscape contract. A project can be deemed a success if after one year at least 20–50 percent of the area is covered and more than 50–80 percent of the plants have survived.

Developing Performance Criteria

Data collection and evaluation may have to occur over a period of years in order to assess the ultimate success or failure of any wetland creation or restoration project (3-71).

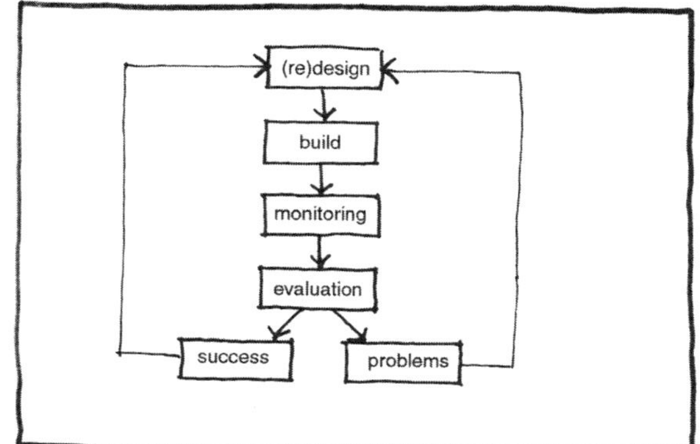

3-71

Improving Design Guidelines

Because wetland creation is a rapidly developing field, and the potential exists for future designs to be vastly improved through incorporation of lessons learnt (and honestly shared) from past projects, production of reports detailing strengths and weaknesses are extremely valuable (though unfortunately rare).

PROJECT SUCCESS/FAILURE

Failed wetland creation/restoration projects abound. These failures most often result from opaque goal aspirations, poor engineering and ecological design, inadequate background data collection, absence of adaptive management, the erroneous assumption that project termination occurs with planting, and a cavalier process of critical evaluation.

Goal Resolution

Problem: Goals are often undefined or are unrealistically attainable.
Solution: Explicitly focused, quantitatively measurable, and realistically achievable project goals (3-72).

3-72

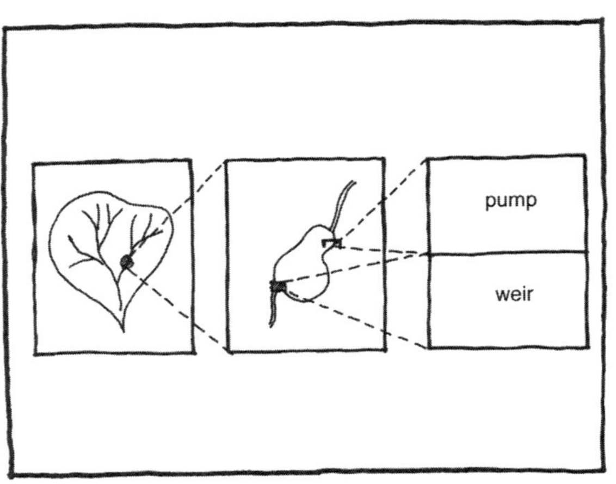

3-73

Hydrologic Imbalances

Problem: Application of incorrect geohydrology due to poorly understood water budget.
Solution: Plan in a watershed context with a detailed assessment of site-particulars and an incorporation of emergency structural measures such as weirs and pumps (3-73).

Plant Mortality

Problem: Plant loss due to choice of wrong species or incorrect planting density in relation to physical-chemical environmental conditions.
Solution: Understand the natural history of the desired species in relation to expected conditions of the created wetland, and produce a detailed and precise planting scheme.

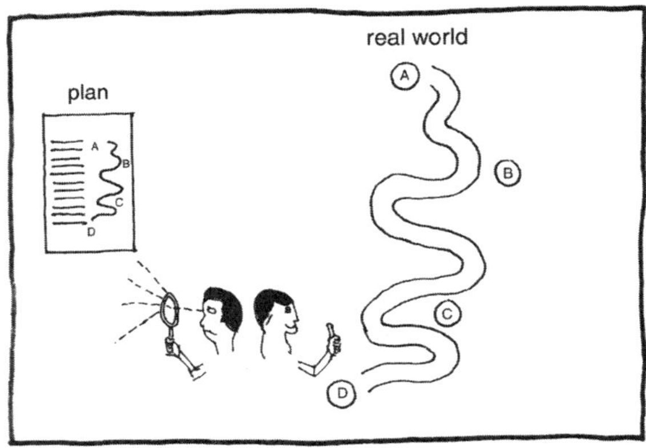

3-74

Nasty Surprises

(a) During project development.

Problem: Unforeseen on-site difficulties arise that necessitate immediate revisions to the initial plans.
Solution: Vigilant supervision by knowledgeable experts is essential for the rapid development of emergency contingency plans (3-74).

(b) After project completion

Problem: Adverse cumulative impacts supersede the functional capacity of the created wetland.
Solution: Plan for adaptive management (i.e., "expect the unexpected") by designing safeguards for worst case scenarios: large buffer zones and reserves for stormwater retention, chemical absorbance, or wildlife refuges (3-75).

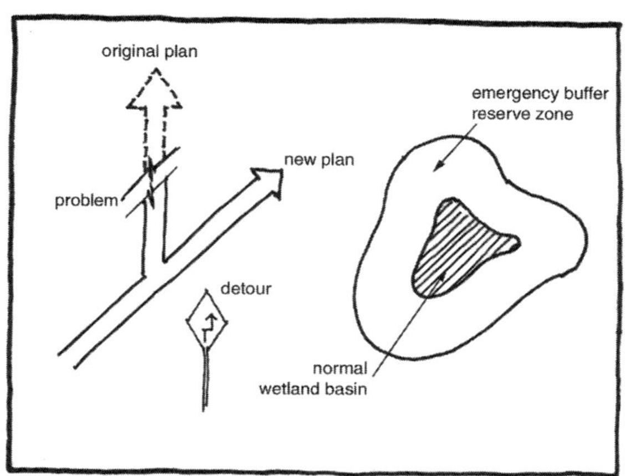

3-75

Evaluation

(a) Initial baseline establishment

Problem: No method employed to evaluate the effectiveness of wetland creation/restoration.
Solution: Make a detailed *a priori* assessment of the function of surrounding/nearby wetlands in addition to itemizing their physical form and static ecological attributes **(3-76)**.

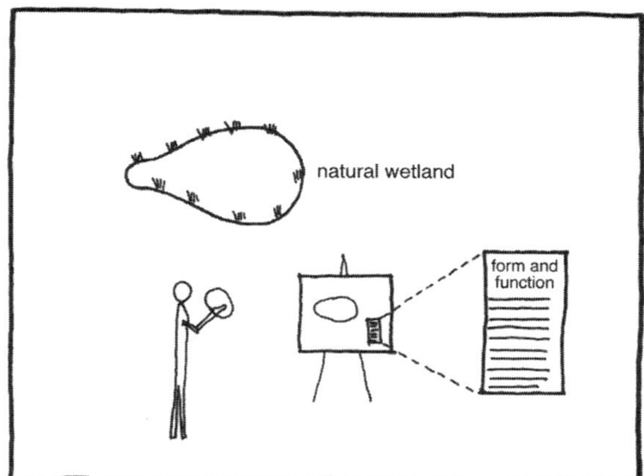

3-76

(b) Regular check-ups

Problem: No project surveillance.
Solution: Adherence to a vigorous and predetermined program for long-term monitoring and maintenance that is supported by a healthy accompanying funding base (3-77).

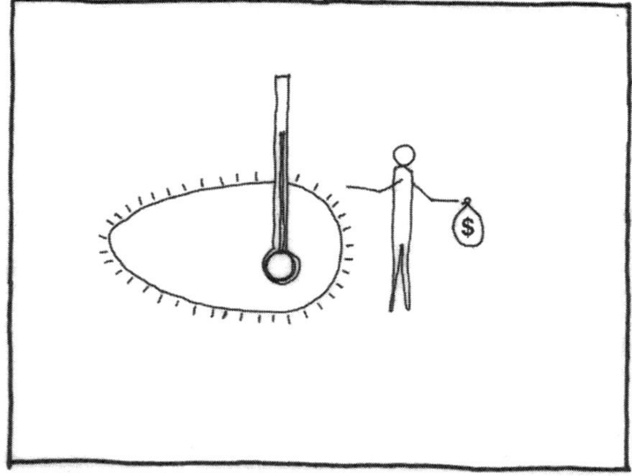

3-77

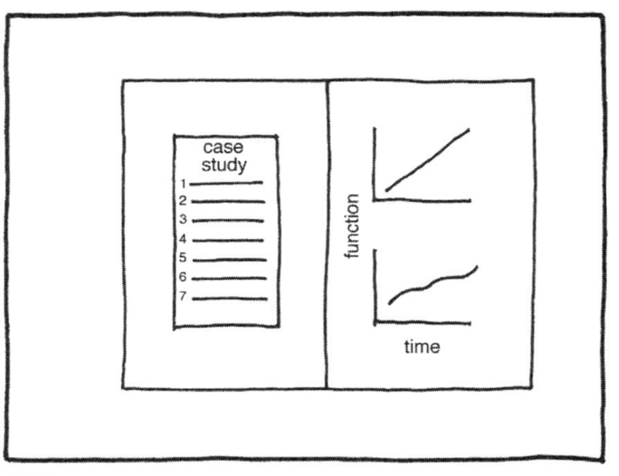

3-78

(c) Final adjudication
Problem: Frequent justification for project success is limited to presentation of photographs showing what is assumed to be healthy plant growth.
Solution: Follow a government-enforced program of comprehensive monitoring, and then present results in a scientifically defensible and reviewed forum in order to ensure a higher standard of success (3-78).

Education

Recognition of the need to preserve undamaged wetlands as well as to restore those that have already been degraded will be fostered by facilitating public access to wetlands having interpretive programs.

SITE ASSESSMENT
Mapping should be undertaken to identify those areas of particular ecological or cultural interest that will be incorporated into the interpretative trail network; particular environmental sensitivity or threat (e.g., flooding hazards) that the public should be kept away from; and locations best suited for constructing other educational facilities.

3-79

ACCESS
Wise and sensitive placement of trails, observation towers, wildlife viewing blinds and platforms, parking lots, boat landings, picnic areas, and especially boardwalks can bring people into close relationship with wetlands (3-79).

INTERPRETIVE SIGNAGE

Although often underfunded, creation of an educational program using illustrations and accompanying text on signs can be an extremely effective way to connect people with wetlands. In situations of limited funding, a trail map can be used in which marks for the most salient features correspond to numbered signposts located throughout the wetland (3-80).

3-80

TEACHING MATERIALS

Information in brochures, reports, slide shows, videos and models, and nonstructural elements such as school programs and guided tours help to inform the public about the goals (e.g., wildlife habitat restoration or stormwater management) of the particular project (3-81).

3-81

Message

Although most interpretive programs convey information about the plants and animals present in the wetland, there is a need to raise public awareness of the multifaceted relationships between humans and wetlands, the dynamic nature of wetlands in terms of hydrological fluctuations and ecological succession, and the wider roles that wetlands play in the watershed (3-82).

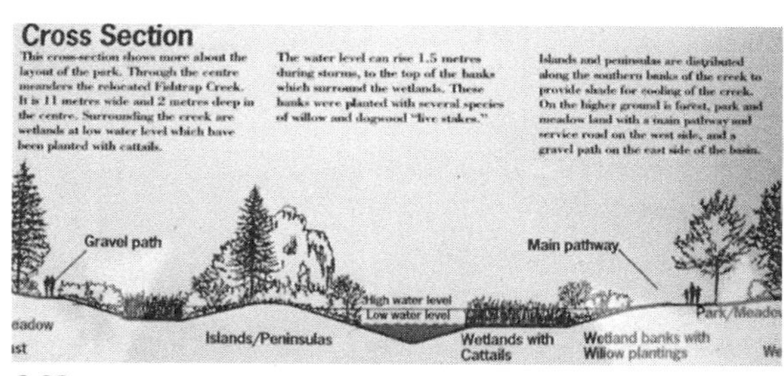

3-82

3-83

Buildings

Kiosks distributed at key sites and a centralized interpretative center are important vehicles for disseminating knowledge (3-83).

3-84

Ecotourism

Wetlands can provide a source of substantial economic benefits to neighboring communities due to their large numbers of visitors (3-84).

Practical Applications

The following illustrations are drawn from over sixty published studies and represent actual examples of real, not theoretical, wetland design projects. What is immediately apparent is the wide range of shapes, sizes and final forms that result from the union of design theory with site pragmatics. The examples are grouped as follows:

- Internal configurations
- Spatial arrangements
- Component networking
- Schematic representations
- Side views

Internal Configurations

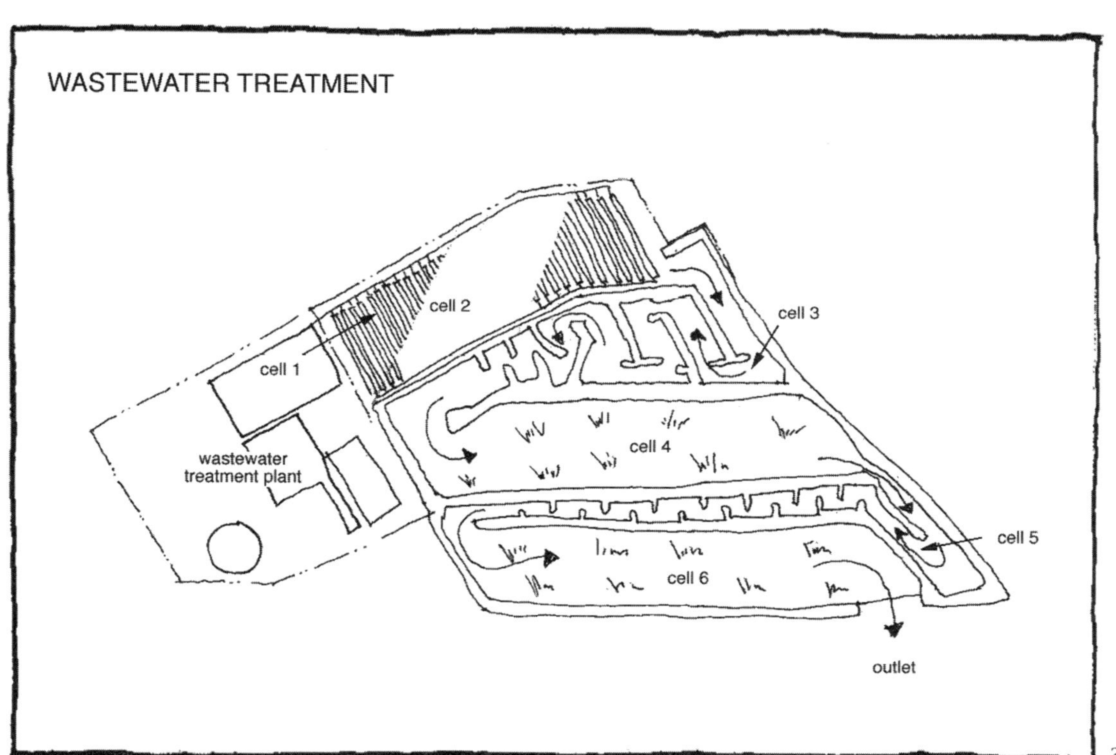

3-85

CONSTRUCTED WETLAND AND
RECREATIONAL POND
connecting pipes
Iris
ROAD
Typha
POND
Scirpus
Phragmites
drain
inlet
Iris
distribution and settlement trench

3-86

distribution ditch
earthen berm
1
meadow
overland
flow
INFLUENT
STRUCTURE
2
marsh
0
800
1600
meters
3
marsh
shallow
pond
deep
pond
5
marsh
deep
pond
weird
4
marsh
outfall
structure
ditch

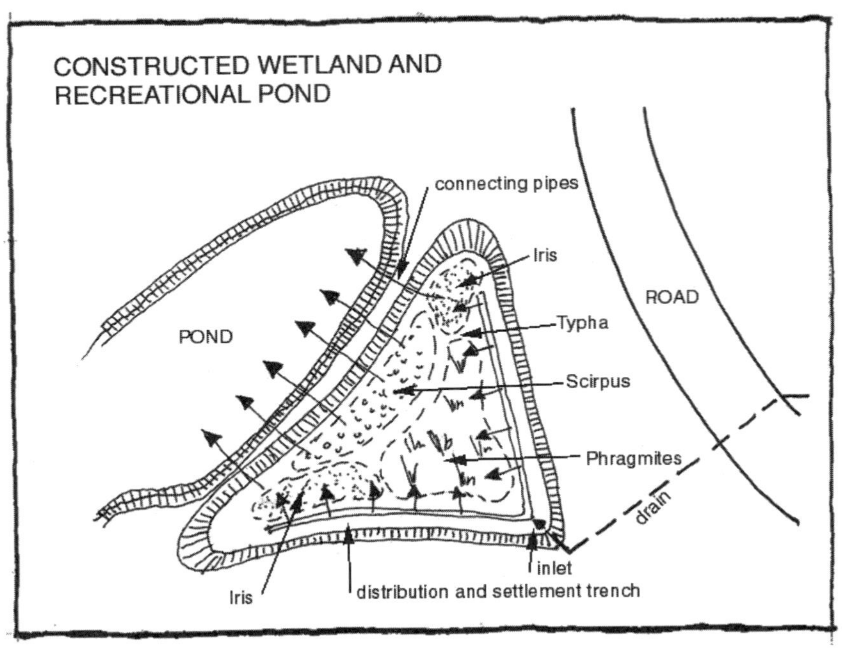

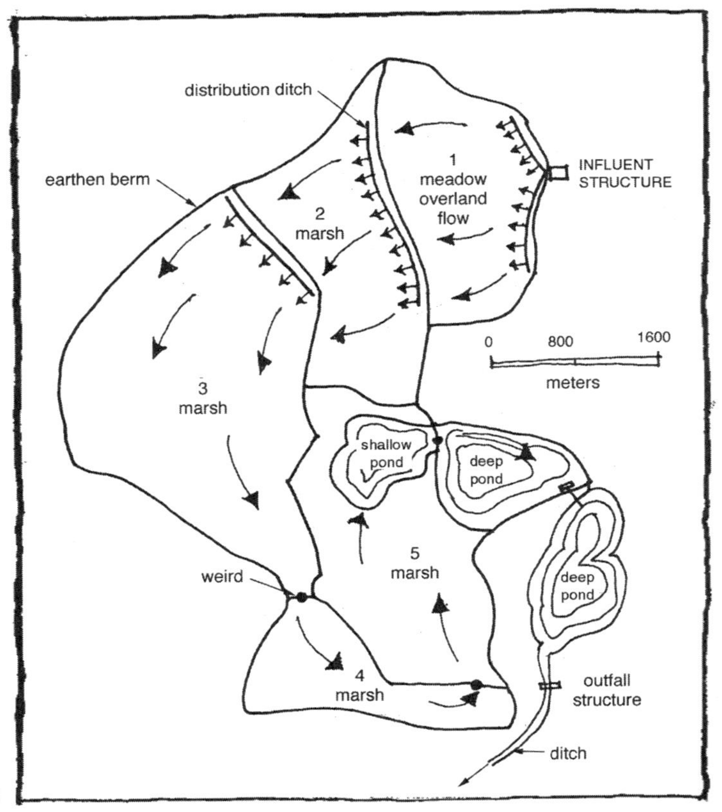

3-87

property line (typ.)
outfall structure
ditch
flow control structure (typ.)
natural hardwood forest
influent
lake
earthen berm (typ.)
LEGEND
cell 1-12: wet prairie
cell 13-15: mixed marsh
cell 16 and 17: hardwood swamp
0 m
600 m
1200 m

3-88

ACID DRAINAGE WETLAND
ash pond
concrete spillway
dike
cell 3 0.21 ha.
final discharge
cell 1 0.30 ha
cell 2 0.42 ha.
concrete spillway
inlet pipe
Dominat vegetation
Cattail
Limestone bed with mushroom compost and cattail
finger dike
scale
0 m
30 m.
clay dike
acid drainage collection ditch

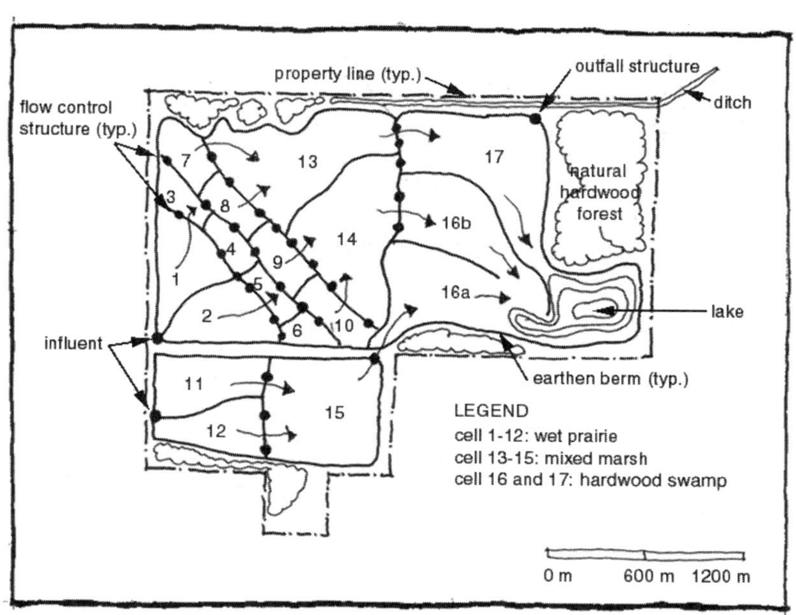

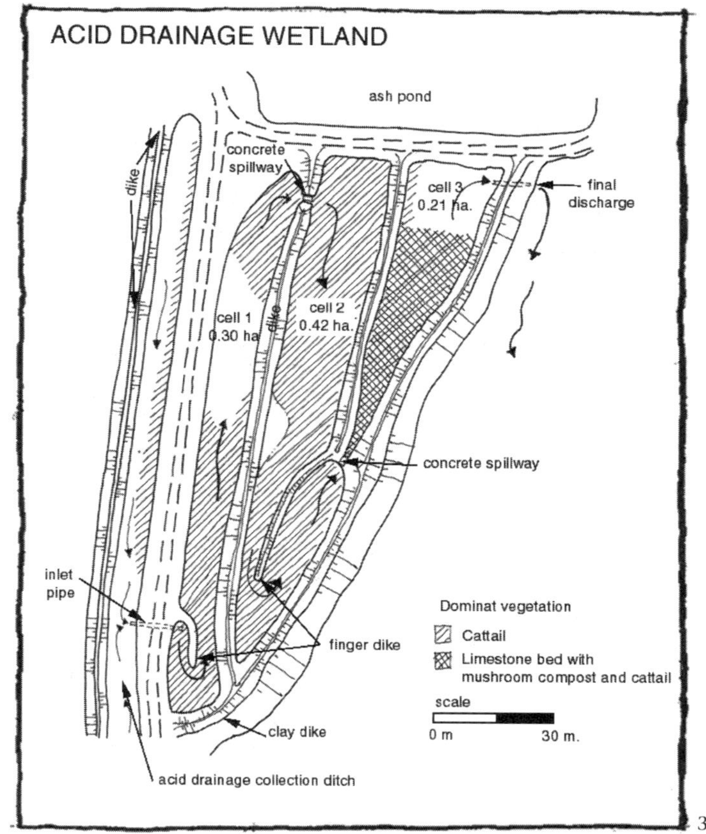

3-89

access road
purap
access road
stream
to storage lagoon
shallow water cells
deep water cells
0 m
130 m

3-90

HEAVY METAL STOCKPILE RUNOFF
limestone bed
flow control culvert
peat mixture
open water pool
open water pool
open water pool
open water pool

3-91

WASTEWATER TREATMENT
to river
flow splitter weir box
influent from primary lagoon
influent trough
50 m.
CELL 1
flow
360 m.
plants-mixture including Iris, Sweet Flag, Cattaila, Arrowhead and Spatterdock
substrate- Native soil
berm
level control structure
influent trough
50 m.
CELL 2
flow
360 m
plants- Woolgrass
substrate- Native soil
outlet
perforated pipes-18 cm.diam.
berm
collector pipes 18 cm.diam.
influent trough
50 m.
CELL 3
flow
360 m
plants- Soft Stem, Buirush
substrate - sized gravel

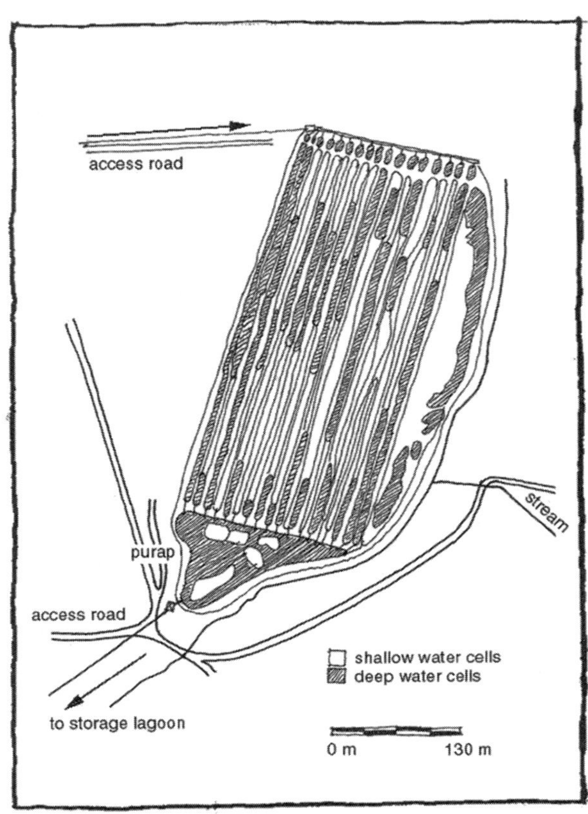

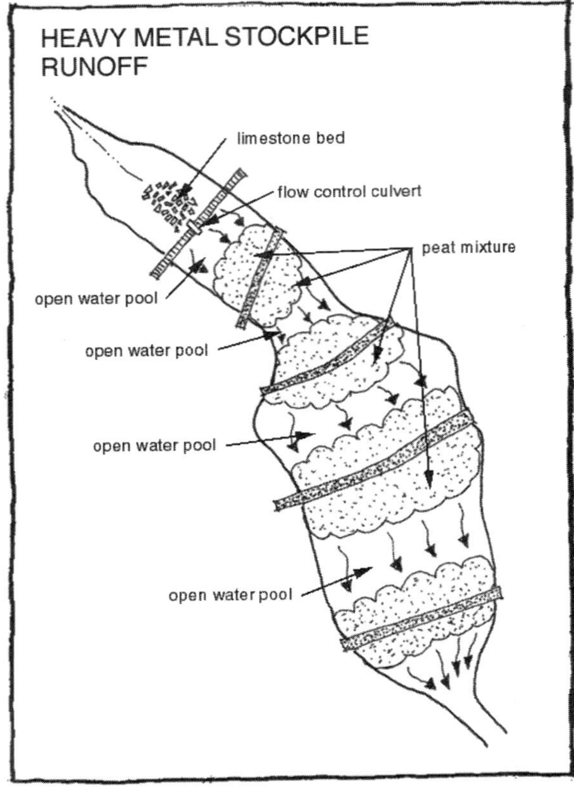

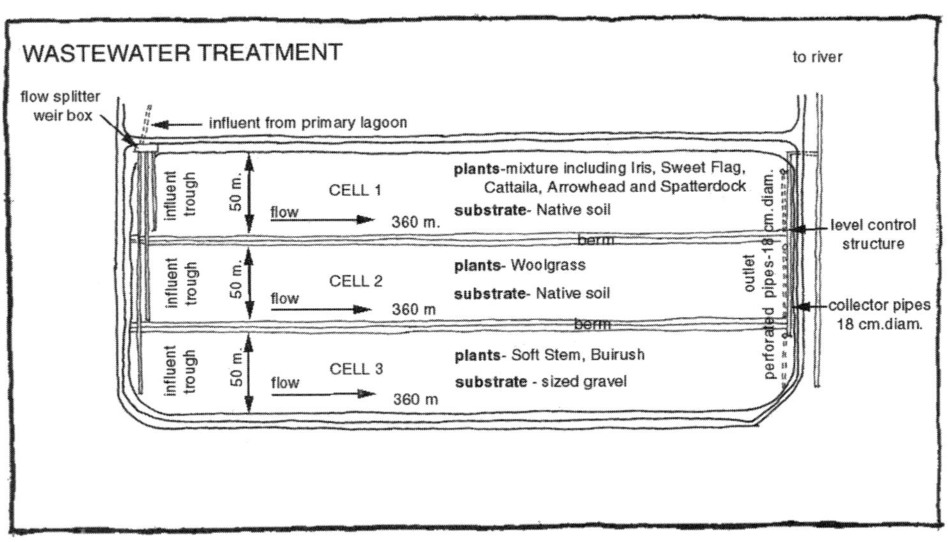

3-92

seep 1
seep area
sandbag
dike
wetland
cement bag
spillway
wetland
wetland
wetland
Major types of
planted vegetation
cattail
bullrush
final discharge
Scale
0 m
30 m

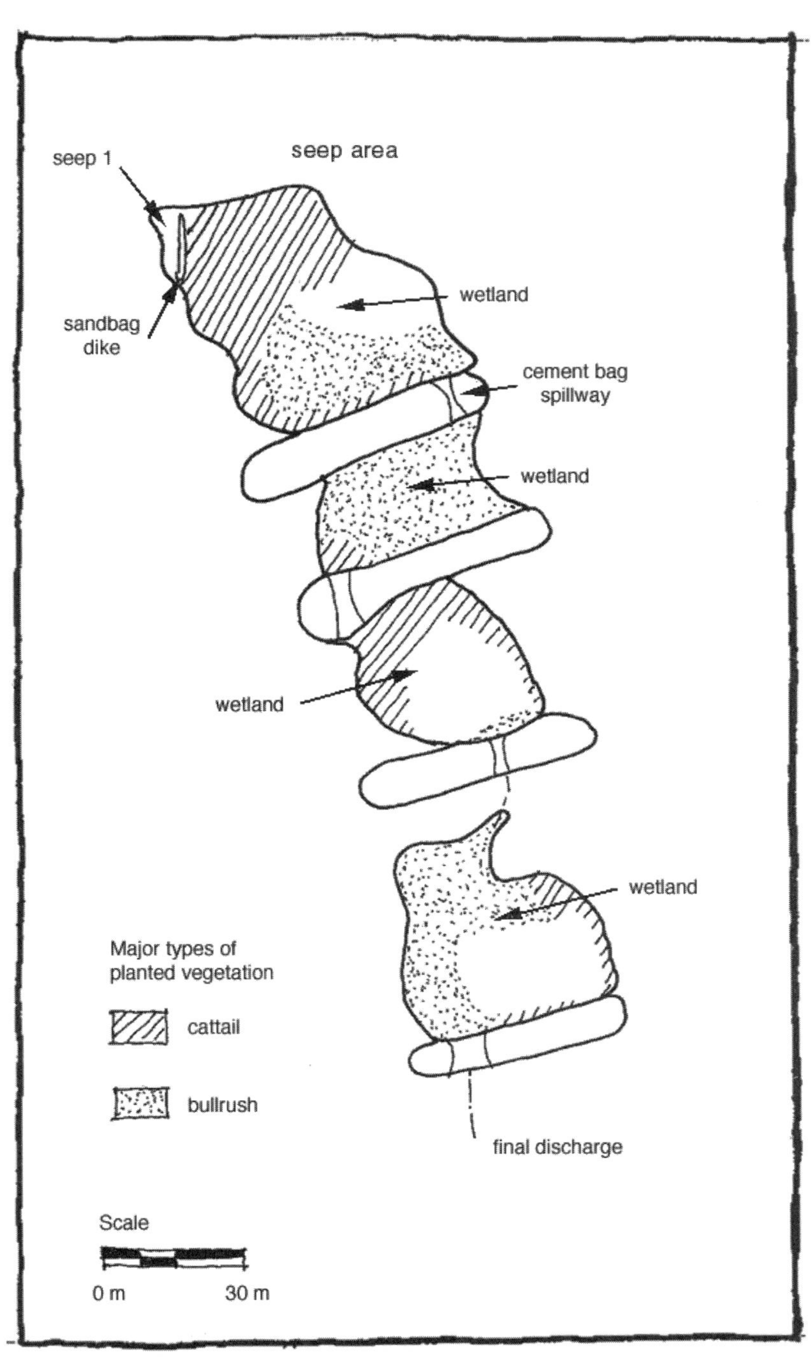

3-93

ACID DRAINAGE TREATMENT
WETLAND
ash pond
seepage inflow
cell 1
dike
pumping
station
coal yard
finger dike
cell 2
coal yard
runoff pond
dike
road
cell 3
weir and final discharge
rip-rap
spillway
dominant vegetation
cattail
spikerush
total treatment area
0.5 ha
scale
0 m.
15
30 m.

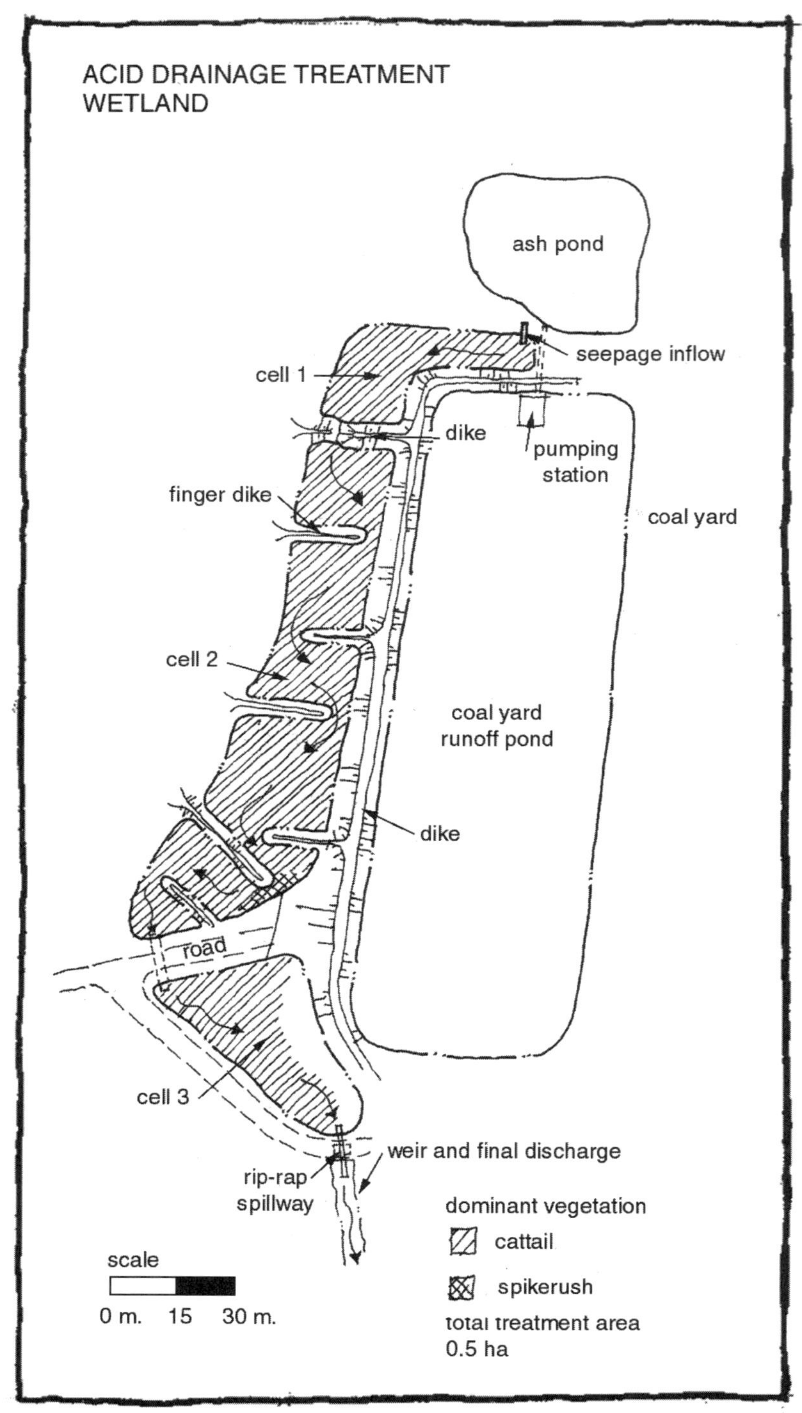

3-94

TREATMENT WETLAND RIVER DIVERSION
outlet
river
outlet
chamber
created
wetland
(1)
weir
berm
settlement
zone
inflow
culverted
river
created
wetland (2)

3-95

SWIMMING BEACH TREATMENT WETLANDS
areation
cascade
uv
swimming area
lake
4
3
2
1
depth 4
1.0 m
Elodea Canadensis
Potamogeton Pectinatus
Vallisneria Americana
0.8 m 3
Alisma Triviale
Elodea Canadensis
Myriophyllum Spicatum
Pontederia Cordata
2
0.5 m
Iris Versicolor
Phragmites Communis
Scirpus Acutus
Typha Latifolia
1.0 m 1
Elchhornia Crassipes
Hydrocharis morsus-ranae

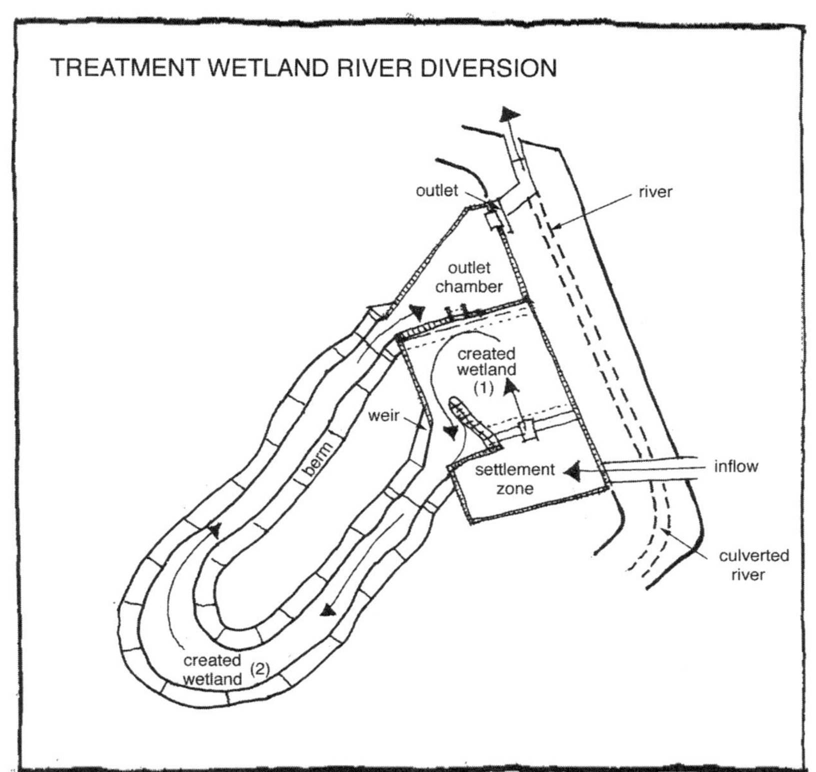

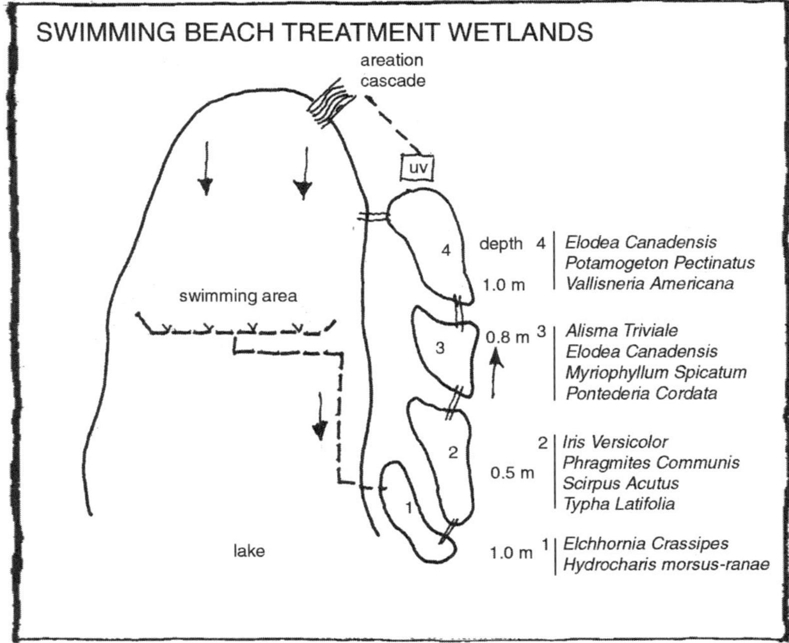

3-96

inner pool
inlet
access road
wetland
outlet
riparian
buffer strip
wetland
scale
0 m
30 m
river

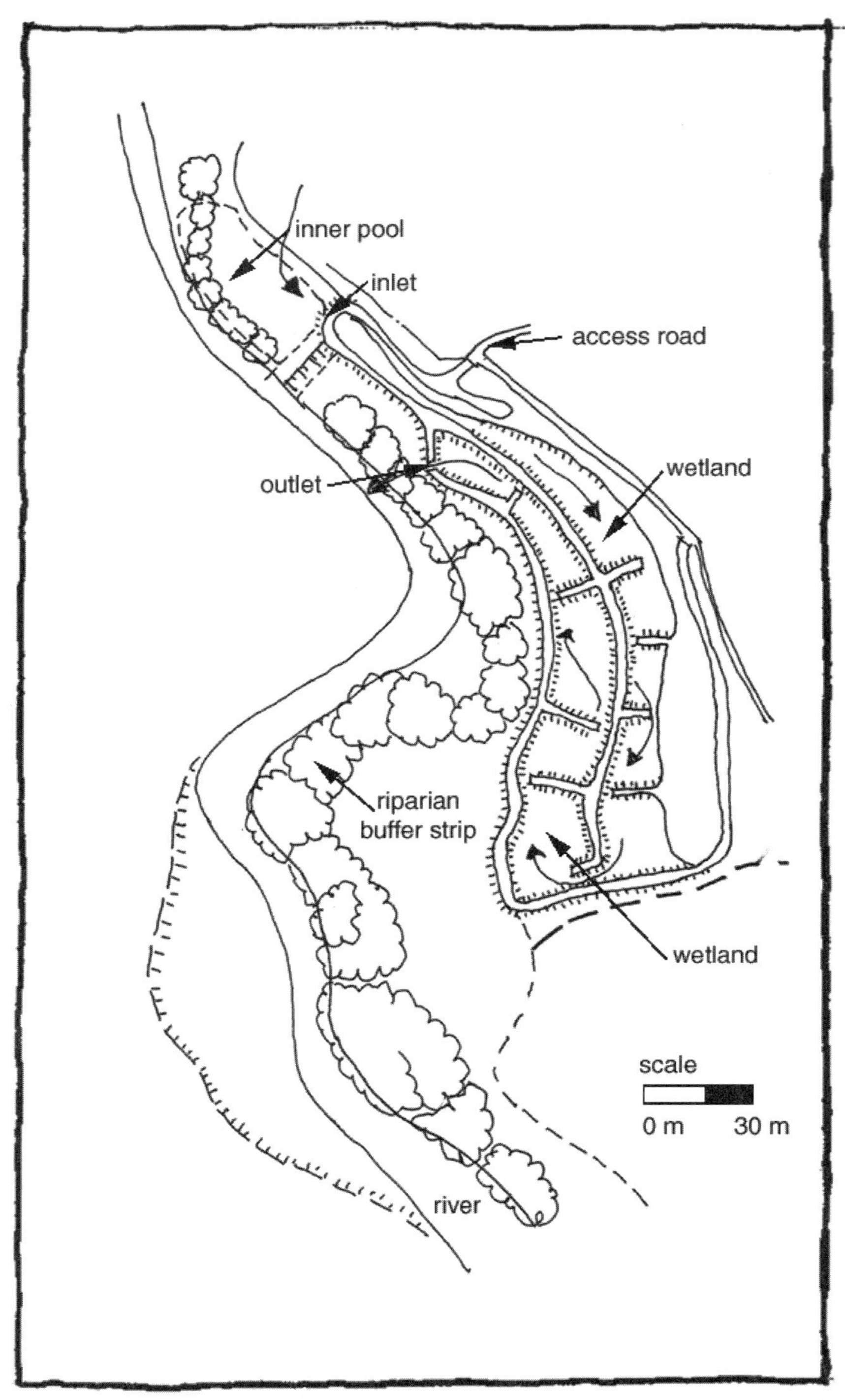

3-97

MINING SITE RUNOFF
existing pond
rip-rap and bullrush spillway
Dominant vegetation
cattail
bullrush
cell 1 0.11 ha
dike
mine spoil runoff
scale
0m. 15m 30m.
cell 2 0.15 ha
rip-rap spillway
dike
rip-rap and spillway
dike
natural wetlands
road
final discharge
3-98

MINE PIT RUNOFF
bay
wetland
creek
pit
seep 1 wetland
stockpiles
wetlands
3-99

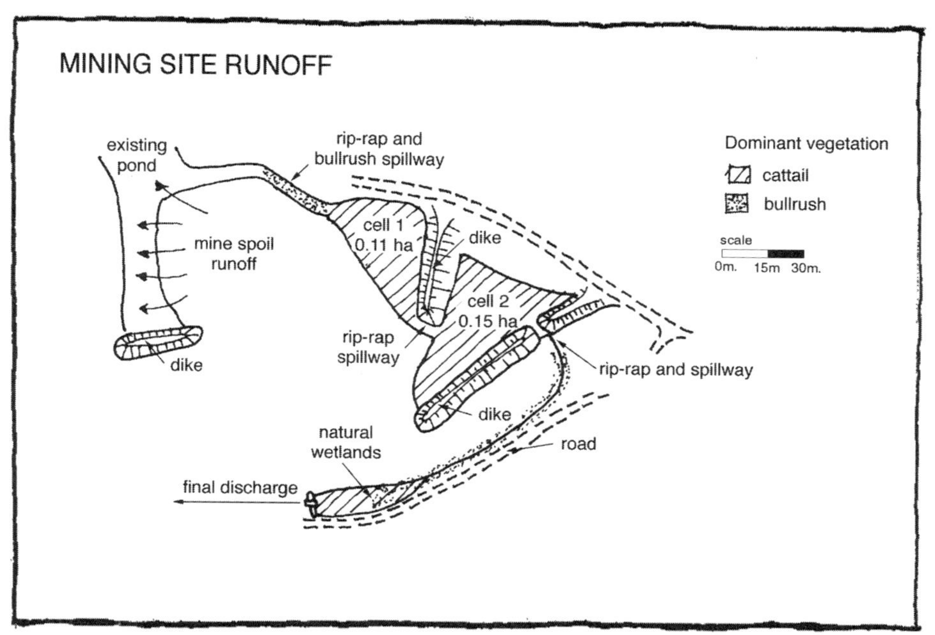

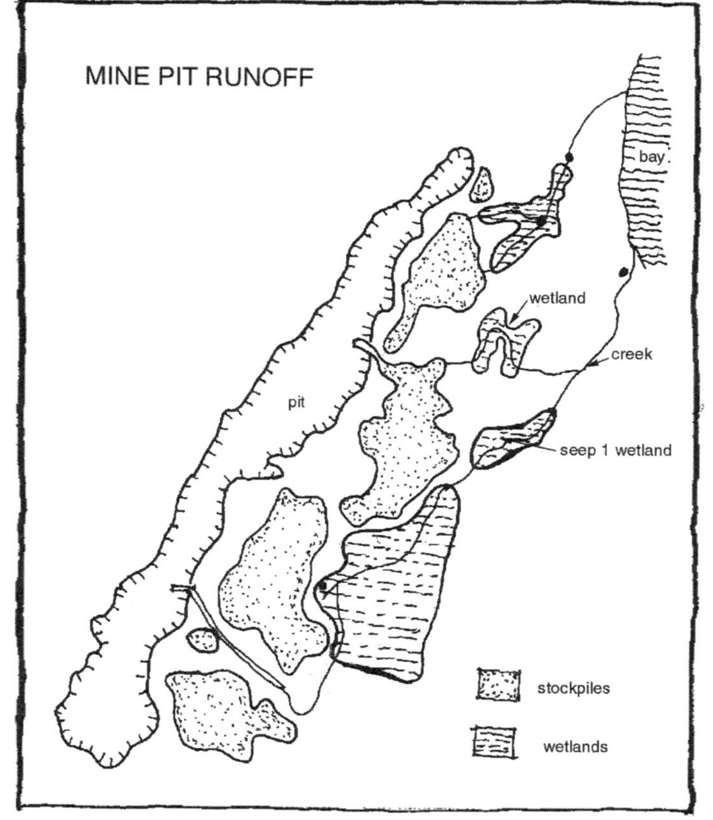

INTEGRATED CREATED AND NATURAL WETLANDS
FOR WASTEWATER TREATMENT
collection and redistribution system
overland flow distribution
discharge to stream
berm
natural "jurisdictional" wetland
road
natural "isolated" wetland
created wetland
created wetland

3-100

outlet
structure
outlet pipe
earthen spillway
embankment
wetlands
area
open
channel
0 30 60
meters
silt
trap
embankment
pipe
pipe
inlet structure

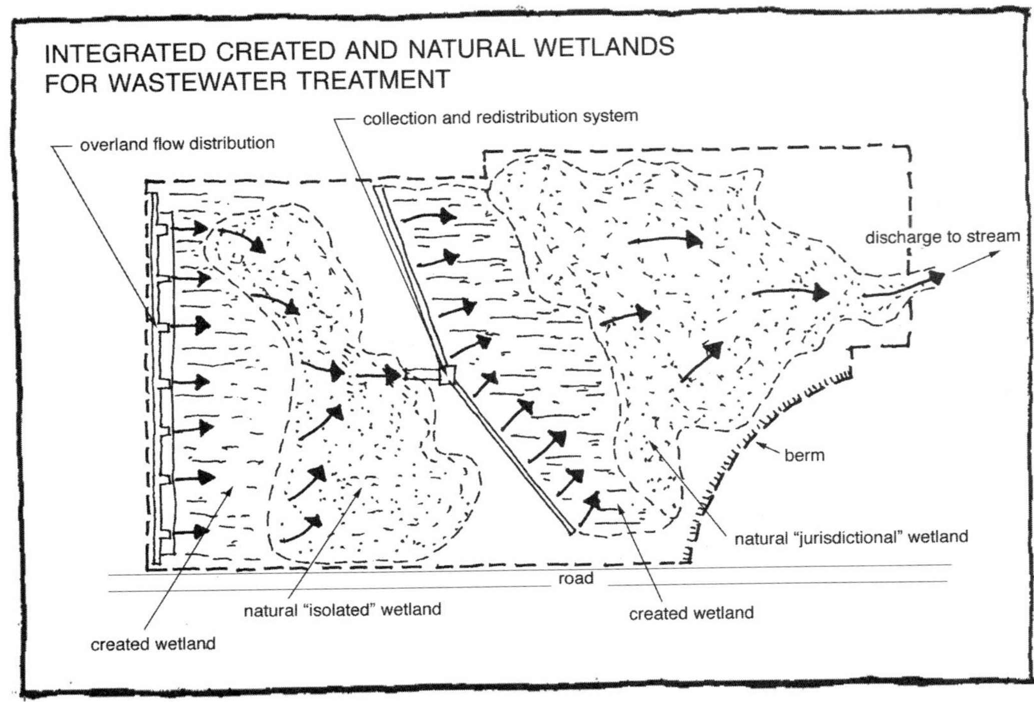

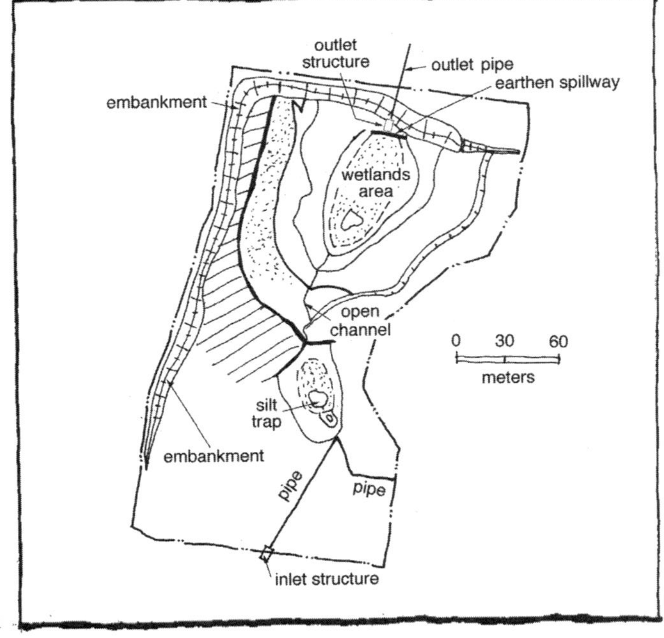

3-101

WASTEWATER
TREATMENT FACILITIES
to lake
river
ARTIFICIAL
MARSH
stoplog weir
spillway
treatment cells
gated culverts
scale
stop log weir
overflow
0m 50m 100m
triple box culvert
pump down
system
concrete spillway
gravity-flow
intermittent filter
diversion
conduit
FILTRATION
IMPOUNDMENT
heavy sediment
collection basin
inflow

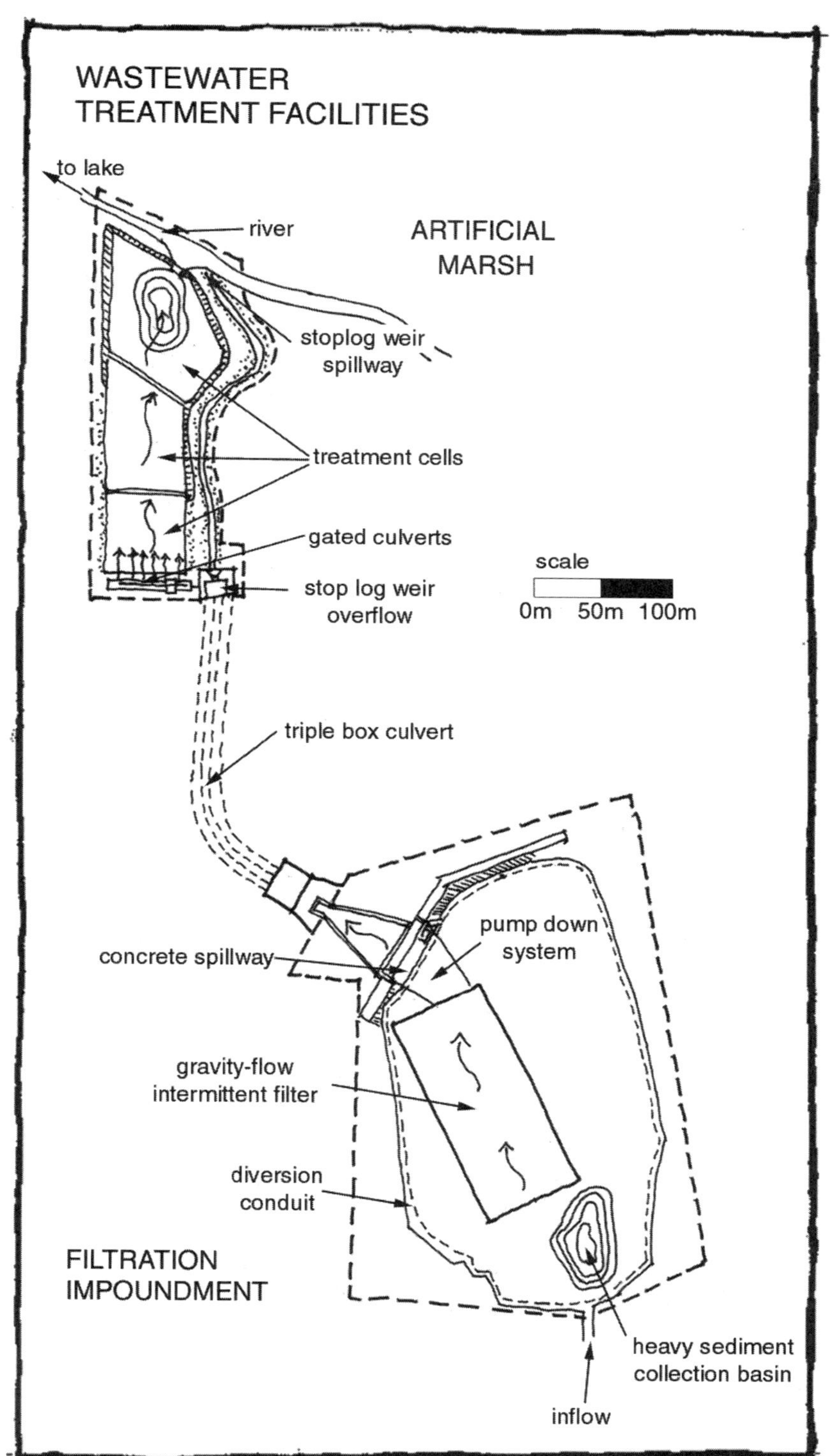

3-102

CREATED MULTIFUNCTIONAL WETLAND COMPLEX
controlled
discharge
mixed hardwood swamp
with mixed emergents
mixed emergent
marsh
mixed emergent
marsh
cattails, rushes, etc.
pond
mixed hardwood swamp
with mixed emergents
mixed emergent
marsh
cattails, rushes,
etc.
LEGEND
earth berm
road
boardwalk
waterflow

3-103

ARTIFICIAL WETLANDS TREATMENT SYSTEM
MARSH-POND-MEADOW-TYPE
aeration cascade
100 m.
meadow A
substrate-limestone
plants-Sedge-Rush or Iris
20 m.
meadow B
substrate-native topsoil
plants-rood canary grass
70 m.
pond A
30 m.
pond B
MEADOW
POND
130 m.
substrate - limestone
plants - buirush
cell A1
13 m.
berm
substrate - limestone
plants - buirush
cell A2
weir
substrate - native topsoil
plants - cattails
cell B1
substrate - native topsoil
plants - cattails
cell B2
berm
MARSH
wastewater
treatment plant
existing effluent
exisiting 8" sewer
plant lift
station

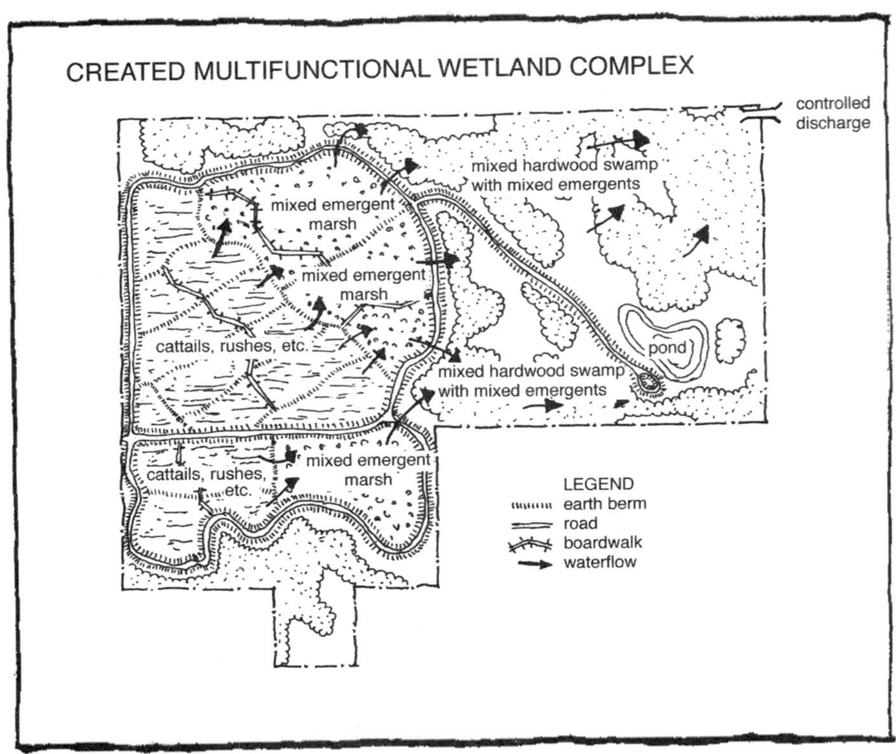

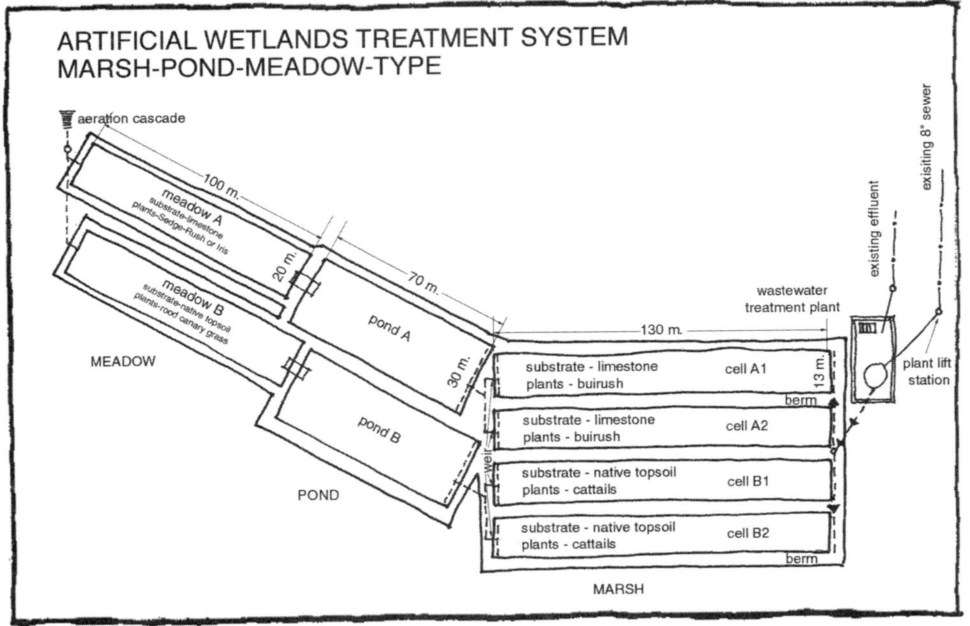

3-104

CONSTRUCTED WETLANDS
RECLAIMED ASH DISPOSAL AREA
limestone drain
OXIDATION POND
2 - 3 m. deep
earth dike
CELL 2
marsh
marsh - pond
dike
marsh
deep
pond
CELL 3
discharge to stream

3-105

CONSTRUCTED WETLANDS
CELL 1
CELL 2
CELL 3
deep
marsh
LIMESTONE DRAIN
shallow
marsh
marsh
pond
CELL 4
dike
spillway
polishing
pond
1.2 m. deep
discharge
underground flow

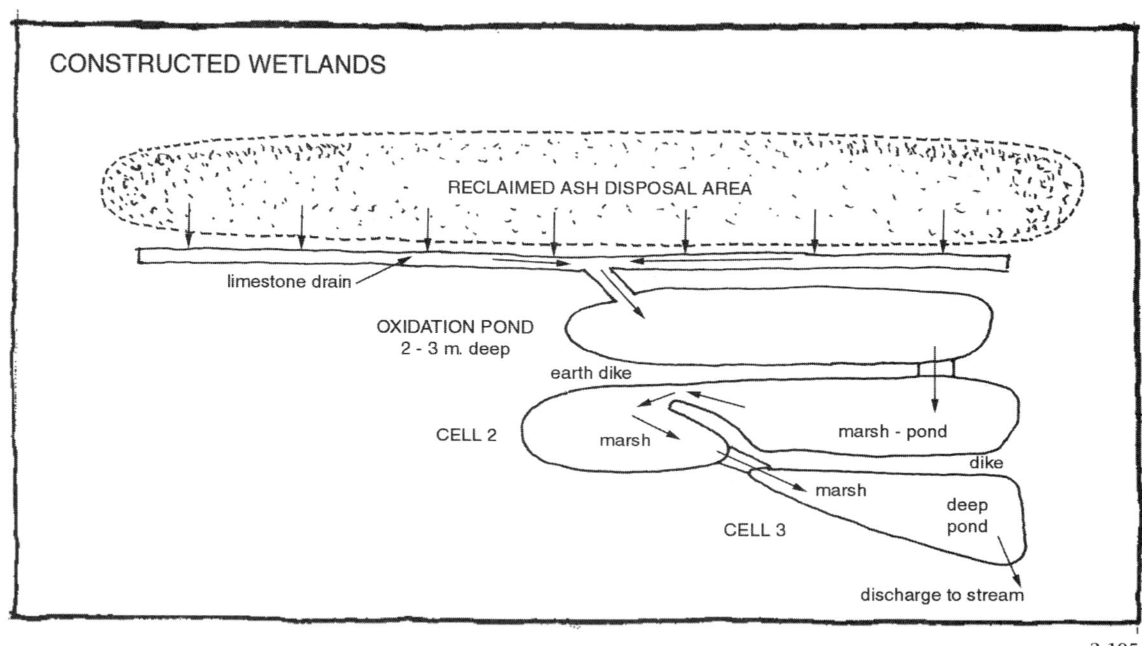

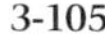

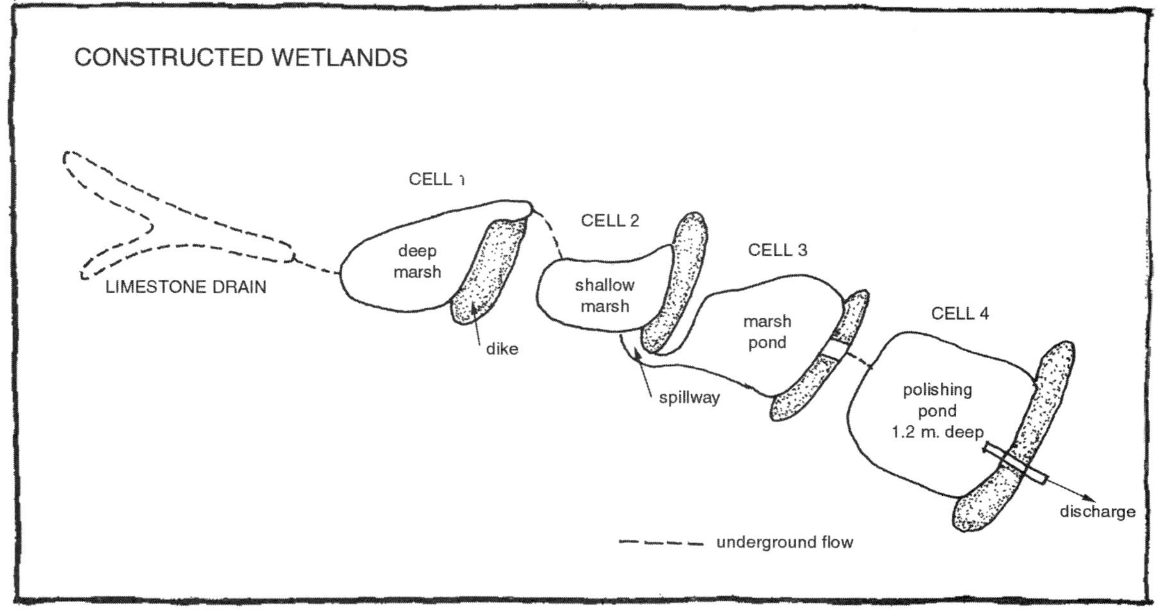

3-106

Schematic Representations

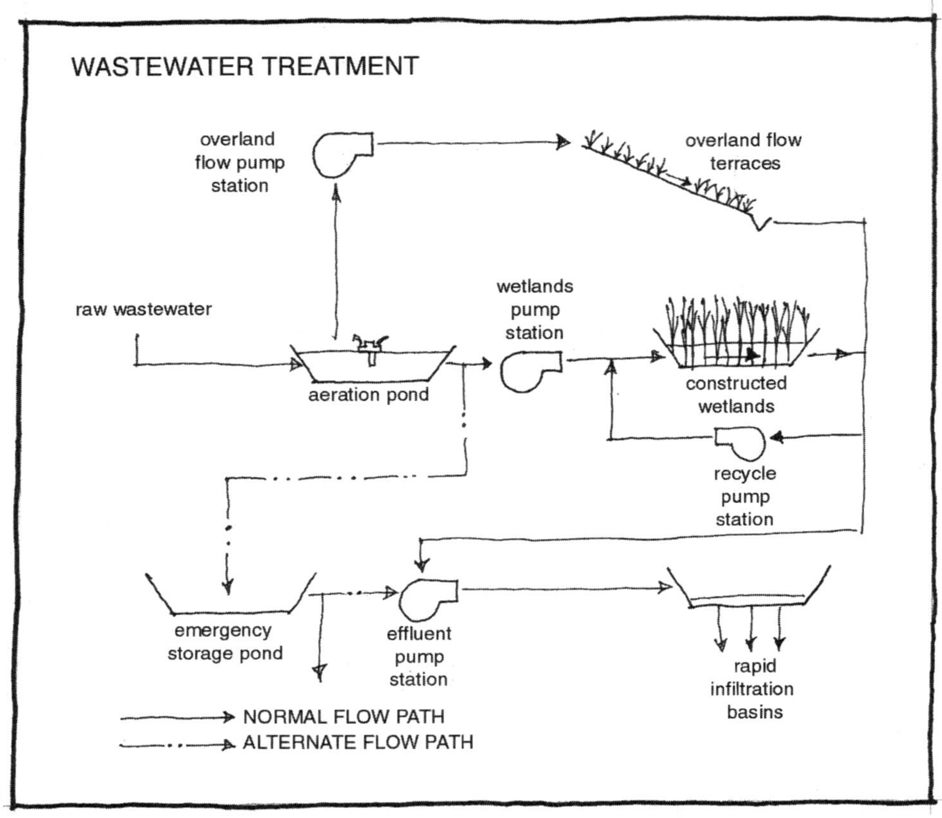

3-107

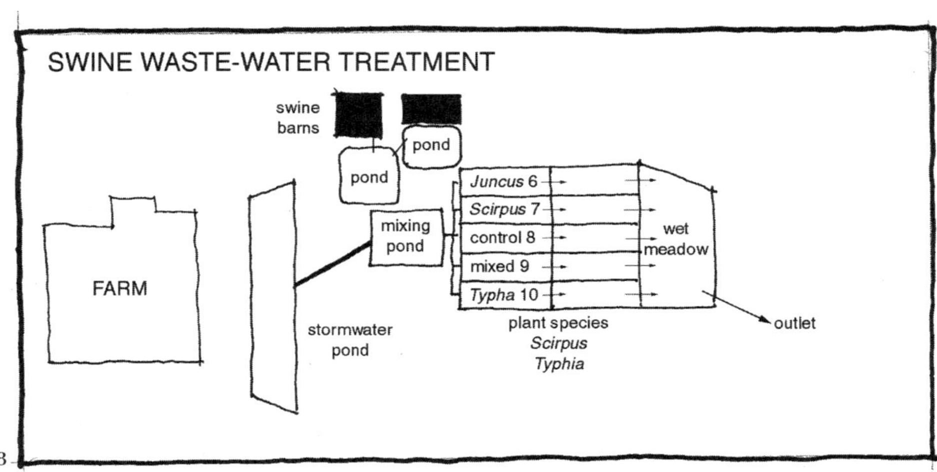

3-108

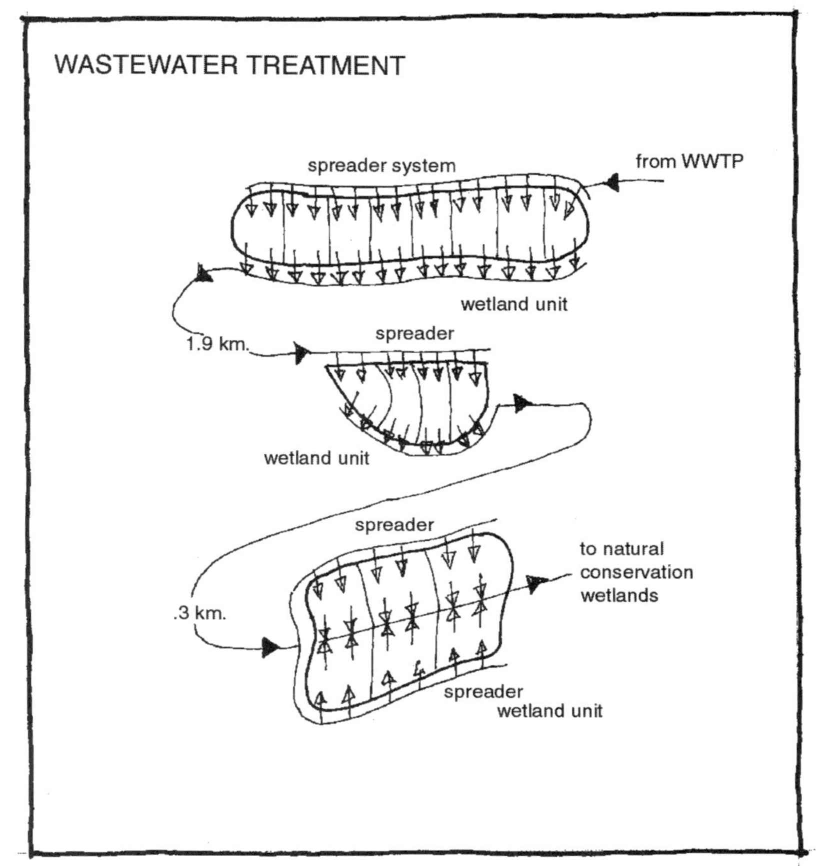

3-109

FISH HATCHERY TREATMENT SYSTEM

backwash water
settling tank
engineered marsh
fingerling rearing tanks
fish-holding tanks
sludge to compost
to ground
engineered marsh
triangle filters
permeable wall of oyster shells
costal zone
biofilter
flitered water
settling pond

3-110

INFLUENT
day 1
1st stage
Phragmites australis
5 m x 1.5 m = 7.5 sqm.
2nd stage
Phragmites australis
5 m. x 1.5 m = 7.5 sqm.
day 2 - 4
3rd stage
Scirpus lacustris
4 m. x 1.5 m. = 6 sqm.
4th stage
Scirpus lacustris
4 m. x 1.5 m = 6 sqm.
5th stage
Iris pseudacorus
4 m. x 1.5 m = 6 sqm.
effluent

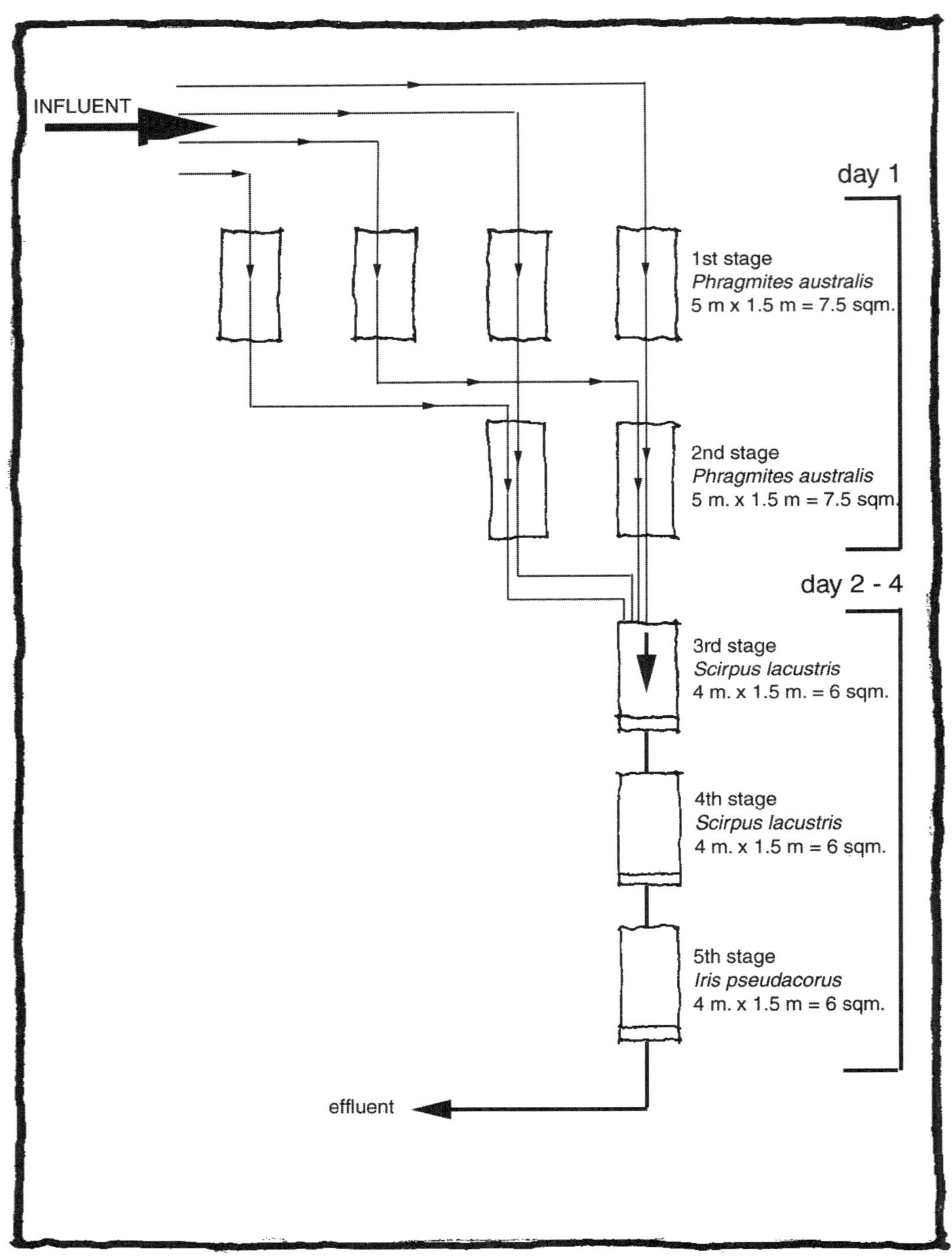

3-111

DOMESTIC SEWAGE TREATMENT
settlement tank
6 X 8 sqm.
Phragmites
beds
STAGE I
3 x 5 sqm.
Schoena plectus
Iris
Phragmites
STAGE II
small settlement tank
recycle
cascade aeration device
Iris
8 sqm. bed
Acorus
Carex
Sparganum
Shoeno-plectus
20 sqm. bed
STAGE V
pond
Tyha
STAGE IV
90 sqm. pond

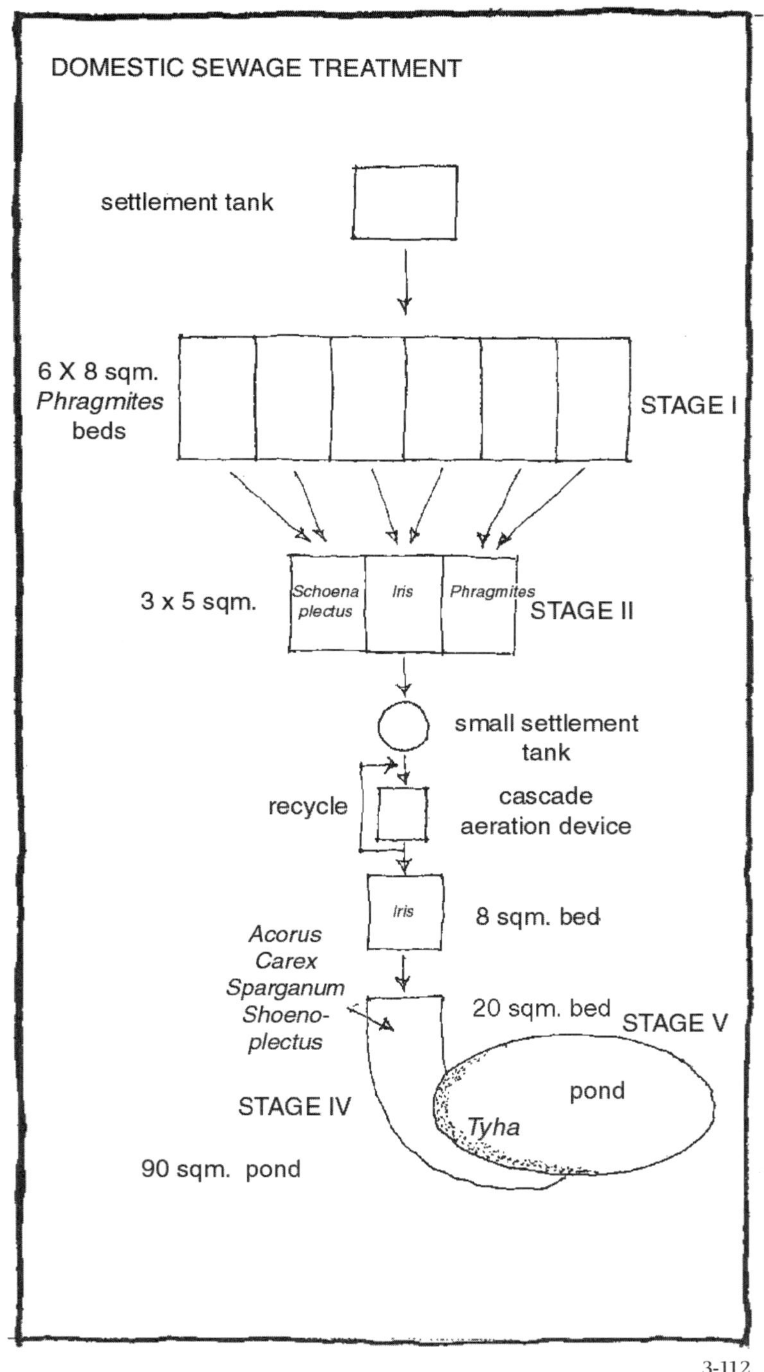

3-112

Side Views

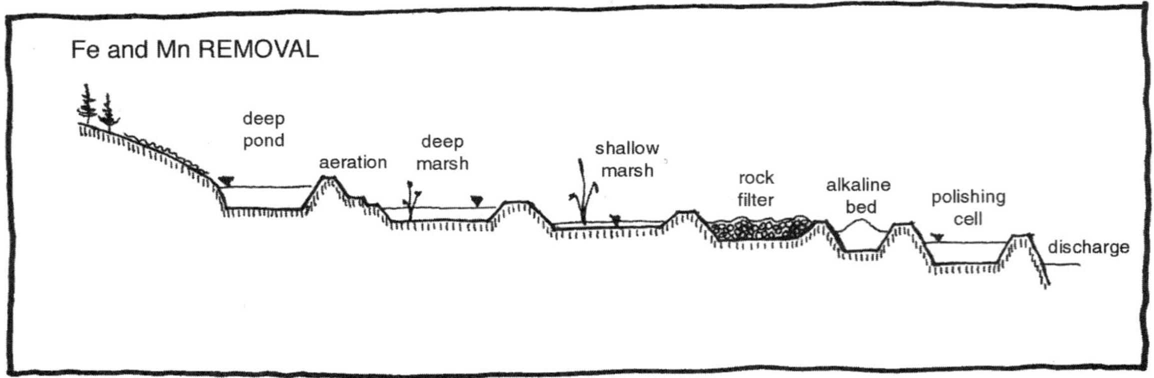

3-113

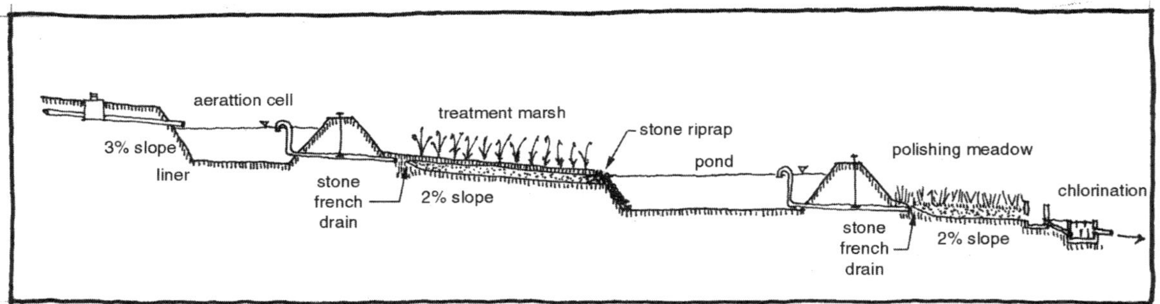

3-114

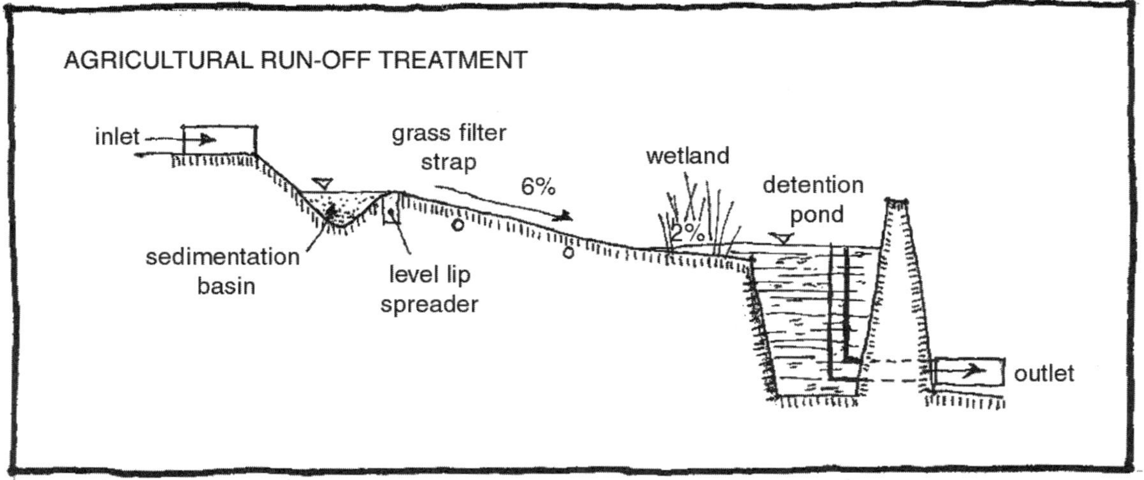

3-115

WASTEWATER TREATMENT
water inflow
sedimentation basin
wetland
final detention pond
water flow
sedimentation basin
overland flow
final detention pond
cell 1
cell 2
cell 3
cell 4

3-116

SMALL-SCALE WASTE TREATMENT
PRODUCTION PRETREATMENT
TREATMENT
DISPOSAL
plastic lined trench
buirushes in gravel
septic tank
wet well with sump pump
ground water

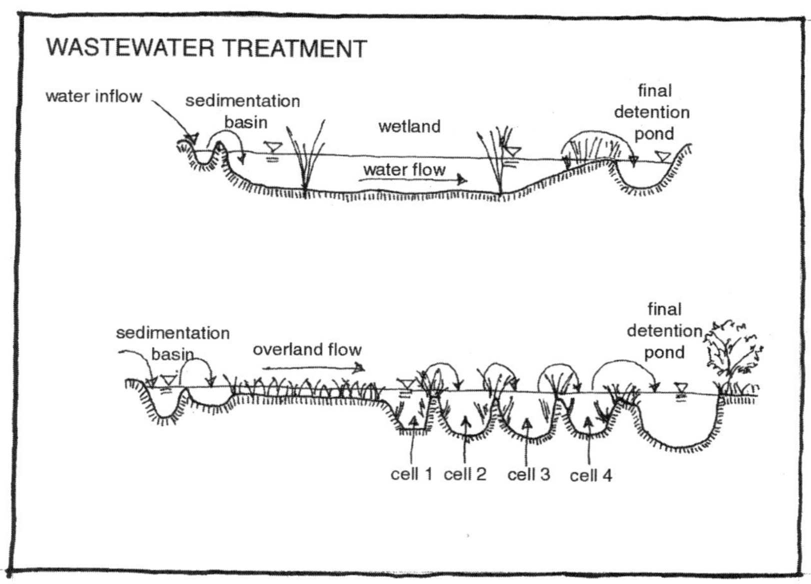

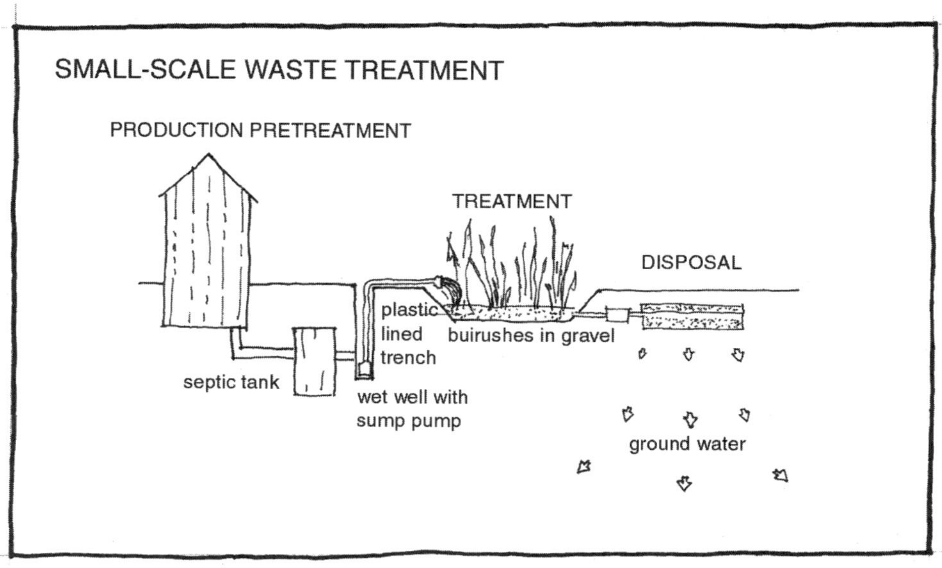

3-117

Case Studies

The following seventeen projects are examples of visionary wetland creation. Each demonstrates effective achievement of the multiple objectives of designed function and form listed at the start. Many of these works have won awards. (Virtual visual tours of these wetlands, and many others, comprising thousands of color images, are contained on a series of CD-ROMs available from the author.)

Fishtrap Creek Nature Park

(Abbotsford, British Columbia, Canada, Catherine Berris Associates, Inc.)

- stormwater retention
- aesthetics and recreation
- education

The original goal of this project was simply to design stormwater detention basins to protect flood-prone agricultural lands located downstream from suburban development. City officials were able to be convinced, however, that in the best tradition of Olmsted's Fenway, this particular site offered far greater potential and could become an environmentally sensitive community park to encourage learning about urban ecology, restoration stewardship and local history, all without compromising the wetland's original hydrologic role.

The resulting wetland park includes walkways, a pedestrian bridge, a picnic shelter, lookout points, seating, and interpretive signage. Of great interest are several signs which not only identify ecological features of interest, as is the standard practice, but also inform readers about the actual design process of wetland creation and stormwater management. For example:

> ***Site Plan—South Basin.*** Fishtrap Creek lies on a 23 ha. site. It was developed as a public park and a stormwater detention facility. . . . From here, you can see substantial residential development surrounding the park. Pavement and rooftops cause an increase in the stormwater runoff downstream of the park; this would have caused flooding of agricultural lands and erosion of the creek. To prevent this large detention basins were constructed on this site. These basins collect water during times of heavy rainfall, and release it slowly back into the creek after the storm is over. . . .
>
> One of the primary goals of the park was to protect and restore the fish and wildlife habitat. Another important goal was providing opportunities for people to enjoy and understand the natural setting.

inflow
Picnic Shelter
Observation Tower
Trail
outflow to south basin
FISHTRAP CREEK - North Basin

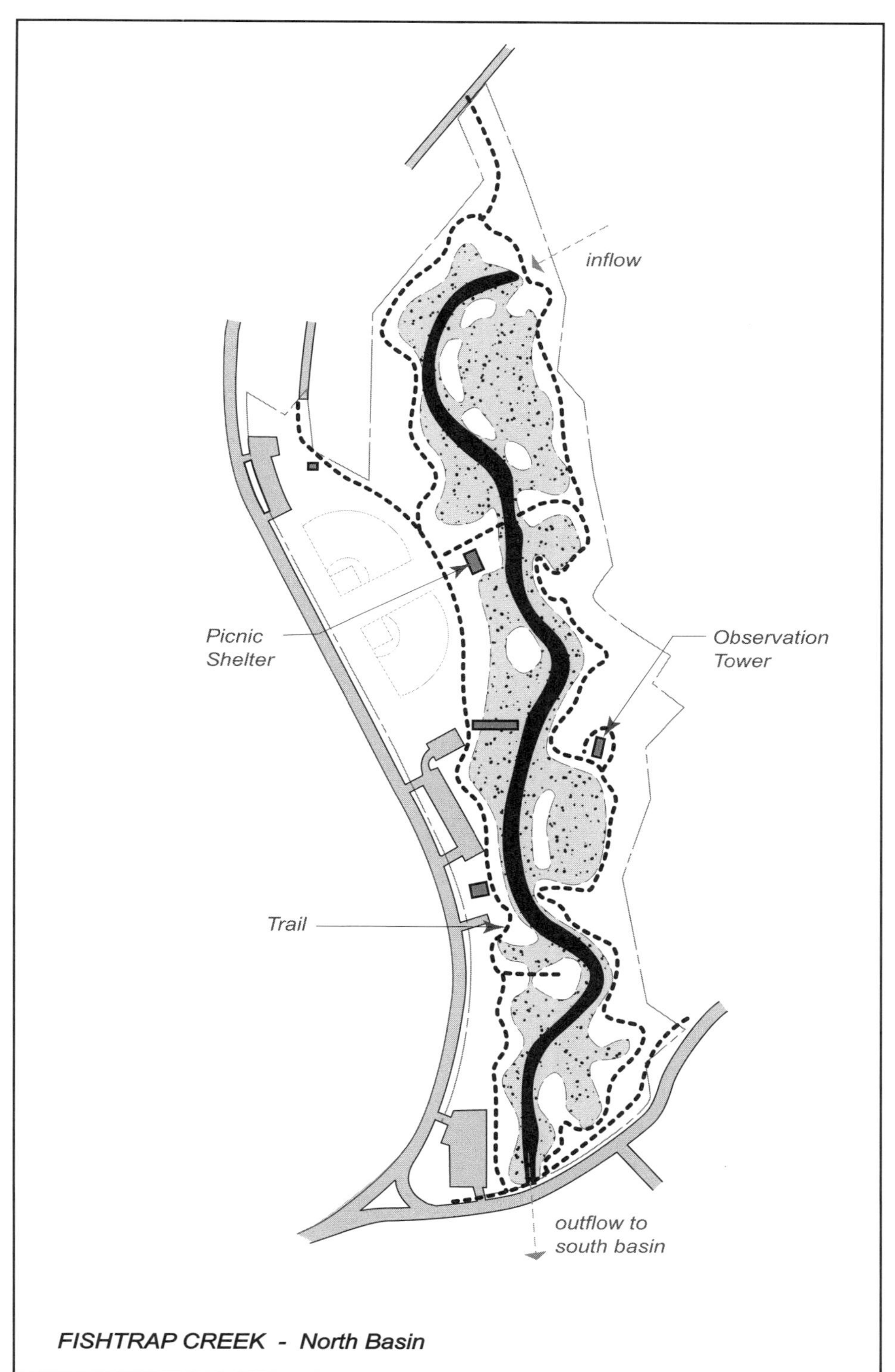

3-118

3-119

The creek meanders through the centre of the site. It supports Cutthroat trout and two endangered fish species. . . . Around the creek, the wetland detention area was planted with cattails. Island and peninsulas provide a habitat away from people. Trees and large shrubs such as Black cottonwood and willows are planted around the wetland. This keeps waters cool enough to support fish and provides an important habitat for wildlife.

Because a large portion of the original wetland and mixed forest had to be excavated for the water bodies, much of what you see has

3-120

been planted using native species. In the south basin, about 700 trees, 15,000 shrubs and groundcover plants, 3,000 riparian wattles and 36,000 cattails were planted. This began the process of reestablishing the wildlife habitat and maintaining the food web which supports insects, birds, fish, and mammals.

Farther from the water, you will see masses of wild roses, berries and other shrubs. You will also notice areas in the park where the grass is mowed. These areas are meadows, which also provide food for a variety of birds and animals. Some of the original forested areas of the site were left undisturbed. In other areas, large-scale planting of Douglas fir, Western red cedar, Red alder, and Bigleaf maple was done to recreate the forest habitat crucial to the wildlife that inhabit the park.

Cross-section—this cross-section shows more about the layout of the park. Through the center meanders the relocated Fishtrap Creek. It is 11 m. wide and 2 m. deep in the center. Surrounding the creek are wetlands at low water level which have been planted with cattails.

The water level can rise 1.5 m. during storms, to the top of the banks which surround the wetlands. These banks were planted with several species of willow and dogwood 'live stakes.'

Islands and peninsulas are distributed along the southern banks of the creek to provide shade for cooling of the creek. On the higher ground is forest, park and meadow land with a main pathway and service road on the west side, and a gravel path on the east side of the basin.

Today, only six years after completion, downstream flooding is abated and endangered species flourish. Lunch-hour joggers from nearby offices weave about children in strollers as they make their way towards the highest lookout tower where, in an already established local ritual, they ascend the steps to tag the historic railway tie beside an interpretive sign.

REFERENCES

ASLA. "Back to Nature: Design Merit Award." *Landscape Architecture* 9/97 (1997): 61–62.

Berris, C. "Fishtrap Creek Nature Park." In *Handbook for Water Sensitive Planning and Design*, edited by R. France. Boca Raton, FL: CRC/Lewis Publishers, in press 2002.

Mooney, P. "Revisiting Fishtrap Creek." *Landscape Architecture* 9/01 (2001): 66–69.

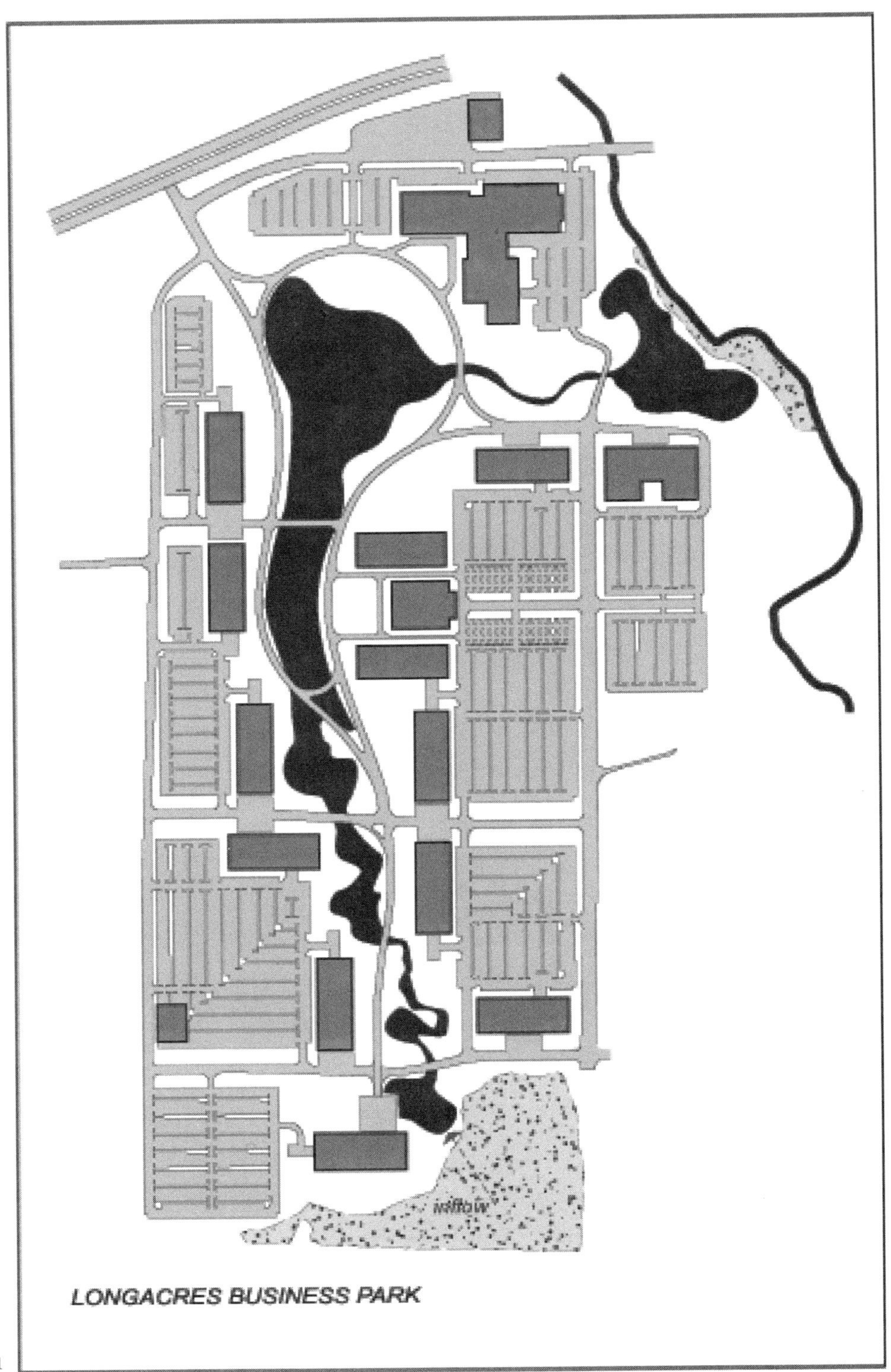

3-121

Longacres Business Park

(Renton, Washington, U.S.A., Peter Walker William Johnson and Partners)

- establishment of site identity
- aesthetics and solace
- wetland restoration and historical preservation
- stormwater contaminant treatment

The site of an abandoned racetrack was selected for a corporate office campus. Now a region of residential sprawl, it was once part of an ancient river that moved through an intermittently flooded forest at the beginning of the century.

The three major project goals were to restore a degraded landscape; protect the new landscape from surrounding development and further deforestation; and sensitively incorporate new offices into the landscape while facilitating access for employees. The overall objective was to create a central wetland park focused around the theme of water movement as the unifying feature of identity for the campus. Secondarily, the designed series of ponds, streams, and wet meadows would function as a natural filtration system for stormwater runoff.

The rigorous process of site analysis included mapping existing patterns of stormwater flow, soil typology and vegetation composition for wetland identification, as well as conducting detailed water quality monitoring. This work determined that the infield of the old racetrack contained a number of existing wetlands from an ancient oxbow lake. A major design objective was to create a new wetland system out of the existing fragments in order to reestablish the original patterns of water flow from a nearby present-day marsh, through the office campus, and on towards the neighboring river. The benefits and disadvantages of five design options were closely examined in terms of their projected effects on traffic routing, stormwater quality and quantity, flood protection, earthworks, utilities, and concordance with wetland regulations.

The design set out to establish a sharp contrast between the natural environment of sinuous curves engaging the movement of water and the man-made order wherein the rigid axial geometry of the reforested areas planted with trees in diagonal rows framed views of Mt. Rainier in the distance.

Stands of native trees were used to represent the lowland forests that once existed on the site. Other construction elements, such as instream installations and path materials, were all derived from natural sources.

Offices are linked to the wetlands by wide gravel paths with a secondary system of bark-covered paths and wooden bridges running parallel to the trees which connect the wetland with upslope areas of the site.

3-122

Because office employees were keenly interested in finding a place along the water or in the forest in which to escape the corporate environment, these secondary paths were designed as secluded dead ends. This also had the added benefit of limiting access to the most ecologically sensitive parts of the park. Log benches along the paths recall the property's history as a forest management area.

3-123

REFERENCES
ASLA. "Corporate Trailblazer: Design Honor Award." *Landscape Architecture* 9/97 (1997): 55–56.

Lee, L.C., and Associates. *An Analysis of the Distribution and Jurisdictional Status of Waters of the United States, Including Wetlands, at Longacres Park, Renton, Washington: Final Report.* Seattle, WA: The Boeing Company, 1991.

Sverdrup Corporation and Peter Walker William Johnson and Partners. *Longacres Office Park Initial Site Development Design Report and Construction Cost Estimate.* Seattle, WA: Boeing Support Services, 1993.

Clark County Wetlands Park

(Las Vegas, Nevada, U.S.A., Design Workshop, Inc.)

- recreation and tourism
- fostering community pride
- education
- wetland restoration

Discharge of stormwater runoff and treated sewage effluent from Las Vegas transformed a 6 mi. (10 km.) desert wash located between the city and Lake Mead into a 2,000 acre (809 ha.) wetland. Continued city growth, however, has seriously degraded the system in several ways. Increased channel erosion has lowered the stream bed by as much as 18 ft. (6 m.), thereby uncoupling flows from the vegetated wetland and decreasing travel time through the wash from eighteen hours in 1980 to less than six hours in 1985. This has led to a decline in water quality in Lake Mead. An accompanying increase in soil salinity has increased the dominance of several aggressive exotics in the riparian zone, leading to a decline in habitat diversity and wildlife abundance. Finally, the area has become a neglected resource used only as a site for trash dumping and off-road vehicular traffic.

The wash rehabilitation project focused on five specific goals:

> (1) Develop recreational and tourism opportunities, based on public needs, that are compatible with the conservation/restoration of the Wash. (2) Create social benefits for the Valley by providing opportunities for area residents to gain a sense of pride and ownership of this park. (3) Create educational opportunities to convey the importance and significance of the Wash through various media. (4) Conserve and restore natural resources by protecting and enhancing the ecological resources of the Las Vegas Wash. And (5) Complete a master plan that will guide the design and development of the Park's recreational facilities and support infrastructure.

Stormwater
Inflow
Las Vegas
CITY
OF
HENDERSON
Interpretive
Center
Wastewater
Treatment
Facility
Wetland
Restoration
Wetland
Restoration
CLARK COUNTY WETLANDS PARK
to Lake Mead

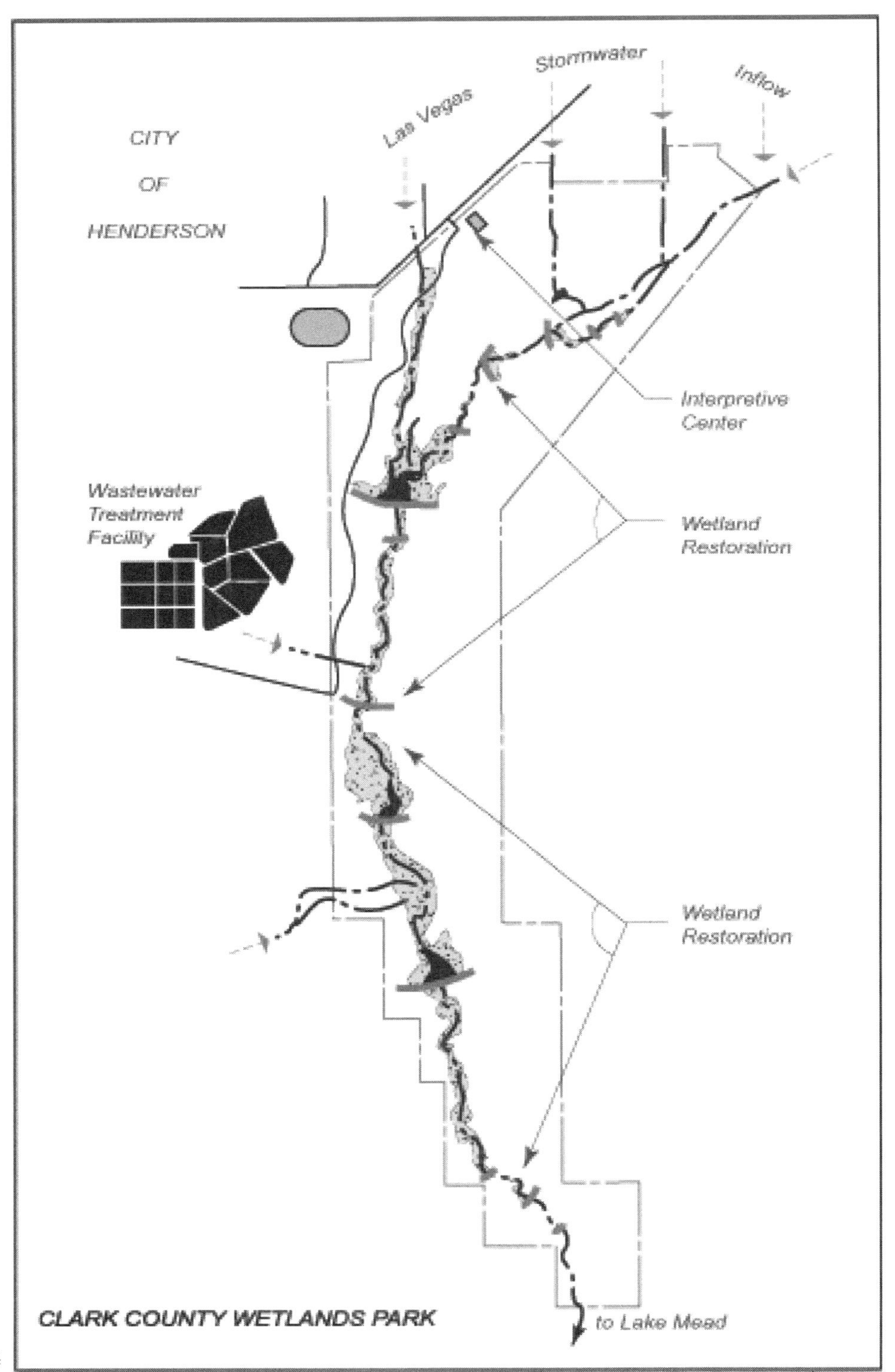

3-124

3-125

A comprehensive GIS-based inventory of biological resources (including rare and endangered species), extent of soil erosion, historical sites, and visual amenities led to the formation of six zones for determining the most appropriate locations for proposed activities. Three categories of activities were examined in this fashion: interpretation and education (trails, signs, observation blinds, platforms and towers), active recreation (horseback riding, jogging and biking), and passive recreation (birdwatching and hiking).

3-126

An extensive public participation process by a large consortium group generated four alternative master plans which differed in their principal focus: conservation (primary purpose is to protect and enhance wildlife habitat), recreation (emphasis on recreational opportunities for people of all abilities), full development (park presented as a significant environmental and recreational resource with major facilities for large numbers of visitors), and the one selected, integration (elements of all alternatives combined).

The final master plan calls for the creation of a park that "has a timeless spirit and reflects the characteristics of the unique landscape." In addition to dealing with the problem of soil erosion/stream channelization and wildlife habitat restoration through wetland creation and bank revegetation, the plan strongly emphasizes aesthetics, particularly in relation to the location and design of the visitor center, trailhead parking areas, road access, picnic facilities, interpretive kiosks, and trail networks.

To ensure project success, a detailed description of step-by-step instructions about how the plan should be implemented was produced. It included considerations of phasing, funding, permitting, and environmental assessments. Finally, a detailed manual on how the resulting park should be managed addressed three distinct, yet interrelated, areas: recreational and visitor operations, erosion control management, and resource policies. In all cases, operational considerations such as remedial maintenance and ongoing design reevaluation were specified to enable adaptive management.

REFERENCES

ASLA. "Betting on a Wetland: Planning and Urban Design Merit Award." *Landscape Architecture* (1997): 68–69.

France, R. "(Stormwater) Leaving Las Vegas." *Landscape Architecture* 8/01 (2001): 38–42.

France, R. "Las Vegas Wash to Clark County Wetlands Park (Las Vegas)." In *Reclaimed! Recovery Processes and Design Practices for Post-Industrial Landscapes*, edited by R. France and N. Kirkwood. In prep., 2002.

Southwest Wetlands Consortium. *Clark County Wetlands Park: Master Plan*. Las Vegas, NV: Clark County Parks and Recreation and Comprehensive Planning, 1995.

Southwest Wetlands Consortium. *Clark County Wetlands Park: Planning Process*. Las Vegas, NV: Clark County Parks and Recreation and Comprehensive Planning, 1995.

Emerald Square Mall

(North Attleborough, Massachusetts, U.S.A., ENSR Inc.)

- stormwater retention
- water quality protection
- recreation

In 1983, in order to build a 700,000 sq. ft. (65,000 sq. m.) shopping center in Attleboro, Massachusetts, a developer proposed to fill 32 acres (13 ha.) of wetland, which would have necessitated the largest wetland fill possible to be permitted in Massachusetts. In exchange, the developer offered to create a 36-acre (14-ha.) wetland out of a nearby gravel pit. The developer was challenged by the EPA, which believed that the fill was not "unavoidable" and that other upland sites had not been adequately explored. The resulting two-year court case received national attention before the developer eventually abandoned the plan.

Finally, while under very close scrutiny from local, state, and federal agencies, a 900,000-sq. ft. (84,000-sq. m.) mall was built at a nearby site by another company in 1986. This scrutiny was due to the new mall being located within a watershed contributing to the regional drinking water supply, and the knowledge that parking lot runoff contains high concentrations of suspended solids, nutrients, trace metals, oil and grease, and deicing salts.

The design had to meet three basic criteria: protection of water quality in a nearby river to the level of the national drinking water standard; maintenance of the existing quantity of flow through the watershed; and control of stormwater runoff peaks and downstream flooding. To achieve these objectives, a complex network of mitigation measures was constructed. It included wet detention ponds, constructed wetland basins, and oil and grease catch collectors, in addition to other non-structural management activities.

The mall parking lot straddles a hydrologic divide with two separate basins. Runoff from the lower basin is directed into two detention ponds located on the mall site and is then routed under the road through a natural wetland channel. From here runoff is diverted into three newly constructed wetlands prior to dispersion into a large wooded swamp which eventually feeds into the river. Due to space constraints, runoff from the upper basin is directed into a detention basin located in an adjacent parcel of land. Here, water is treated by flowing through a series of three created wetlands totaling one acre before being rerouted back under the entire mall parking lot into an existing small pond, from which it is dispersed into a large wooded swamp prior to entry into the river.

Constructed wetlands were operational for over a year before receiv-

1
295
Parking
Wetlands
Lower
Detention
Basins
Upper
Detention
Basins
EMERALD SQUARE MALL

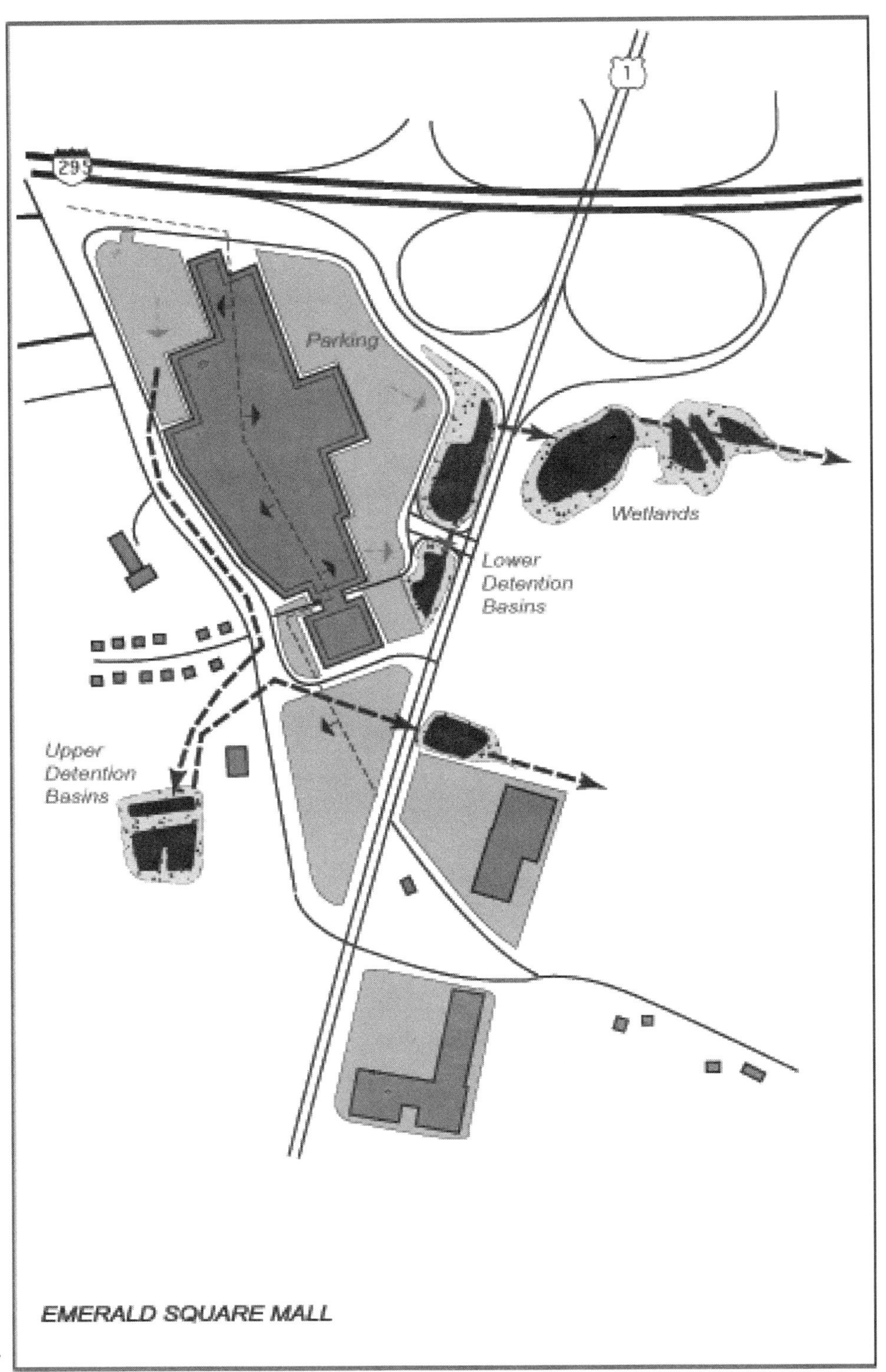

3-127

3-128

ing parking lot runoff and were designed as shallow marshes on organic soil with fringing shrub species to attract wildlife. Stop logs were added to regulate water levels. Planted species include cattails, arrowhead, bulrush and sweet flag. Side slopes were sown with millet for rapid stabilization and reed canary grass for permanent cover. Planted shrub species include dogwood, arrowwood, winterberry, highbush blueberry, sweet pepperbush and willow.

All detention ponds (whose sizes represent about one percent of their respective drainage catchment areas) are designed to attenuate storms up

3-129

to the hundred-year, twenty-four-hour event, and have residence times of one to three weeks for average events. For the upper facility, a berm enclosure was built as an additional safety measure to accommodate overflows onto an adjacent field that provides recreation during dry periods.

Because of the extreme environmental sensitivity of this project, an extensive program of post-construction monitoring is still underway to ensure that contaminant removal efficiencies do not decline below the levels of 50–90 percent presently being achieved.

REFERENCES

Daukas, P., D. Lowry and W. W. Walker. "Design of Wet Detention Basins and Constructed Wetlands for Treatment of Stormwater Runoff from a Regional Shopping Mall in Massachusetts." In *Constructed Wetlands for Wastewater Treatment.*, edited by D.A. Hamer. Boca Raton, FL: Lewis Publishers, 1994. 684–94.

Salvesen, D. "Shoot-out at Sweedens Swamp: the Attleboro Mall case." In *Wetlands: Mitigating and Regulating Development Impacts,* edited by D. Salvesen. Washington, DC: The Urban Land Institute, 1991. 32–33.

Fort Devens

(Ayer, Massachusetts, U.S.A., Carol Johnson Associates Inc.– The Bioengineering Group)

- stream and wildlife habitat restoration
- water quality improvement
- stormwater retention
- aesthetics

The most cost-efficient arrangement of facilities for a medium-security prison built on a decommissioned army base involved relocating an already disturbed stream. Rather than simply move the stream into a traditionally engineered channel, a multi-purpose channel and basin complex was constructed to integrate aesthetics, natural habitats, and stormwater management. In a site constrained by topography, security concerns, and proximity to a national wildlife refuge, a series of three ponds were created along the relocated stream. This design showcased the use of bioengineering techniques for erosion control and water quality management.

The first basin traps most of sand and silt particles from the inflowing stream draining nearby impervious surfaces. Easy vehicular access facilitates maintenance dredging. The second basin is designed as a heavily vegetated wetland for water quality improvement, in which baffles of natural materials help to direct flow and optimize water circulation,

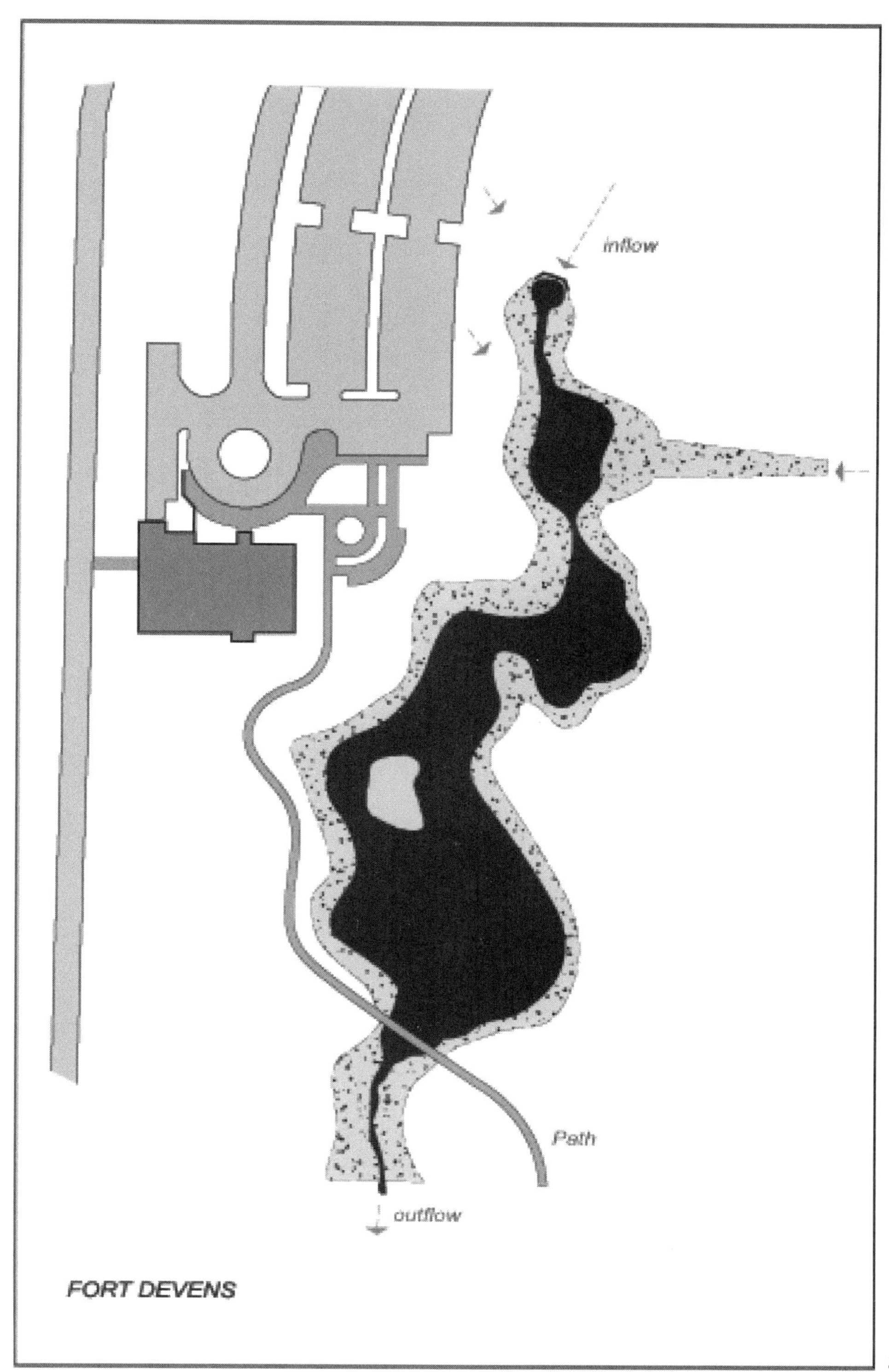

3-130

3-131

thereby increasing residence time and surface area contact for effective contaminant removal. The stream exits over a constructed rocky waterfall designed to aerate the water for increased oxygen levels, as well as to enhance the atmosphere for people sitting on the benches at the nearby overlook. The third basin is a small pond circling an island, both providing wildlife habitat through use of native riparian grasses, sedges, rushes, wildflowers, shrubs, and trees. The entire system is designed to accommodate the hundred-year storm event sized in relation to the watershed surface area, and is hydrologically regulated by a dam and weir system.

Rather than conventional riprap, native shrubs were installed and

3-132

used as the primary method of erosion control along the water's edge. Bioengineering provided added benefits by visually masking the embankments and the dam, thereby attractively blending the earthworks into the surrounding landscape while at the same time serving as a wildlife habitat.

The sequencing of work was carefully managed to optimize erosion control effectiveness. The newly excavated basins acted as temporary sediment traps while the remainder of the site was prepared, and the bioengineered shorelines were planted as the basins were filled.

REFERENCES

Goldsmith, W. "Bioengineering for Enhanced Stormwater Management." *Land and Water* 10 (1997): 28–31.

Thompson, J.W. "Stormwater Unchained." *Landscape Architecture* 8 (99) (1999): 44–51.

Arcata Wastewater Treatment Marsh Facility and Wildlife Sanctuary

(Arcata, California, U.S.A., City of Arcata)

- retrofitted wastewater treatment
- wildlife habitat
- ecological restoration

In 1977, in order to comply with new regulations restricting the discharge of wastewater into enclosed bays and estuaries, the City of Arcata proposed use of a wastewater treatment system which incorporated existing primary sedimentation facilities into newly constructed wetlands. These new wetlands were to be created from former oxidation ponds. The entire site was a degraded urban waterfront.

A detailed pilot study of twelve experimental cells was run for three years to explore the potential of the site for contaminant removal. Sampling consisted of standard chemical and suspended sediment determinations in addition to tracer and disinfection efficiency studies. Following favorable results, several oxidation ponds were converted into treatment wetlands.

In operation, wastewater from the two remaining oxidation ponds is passed into the newly created intermediate treatment marsh before it is routed to the pump station where it is chlorinated and dechlorinated before passing into a series of constructed wetlands which serve as a wildlife sanctuary. Polished effluent from the 31-acre (12-ha.) wildlife enhancement marsh is then redirected into the pump station, where it is treated once again before being discharged into the bay.

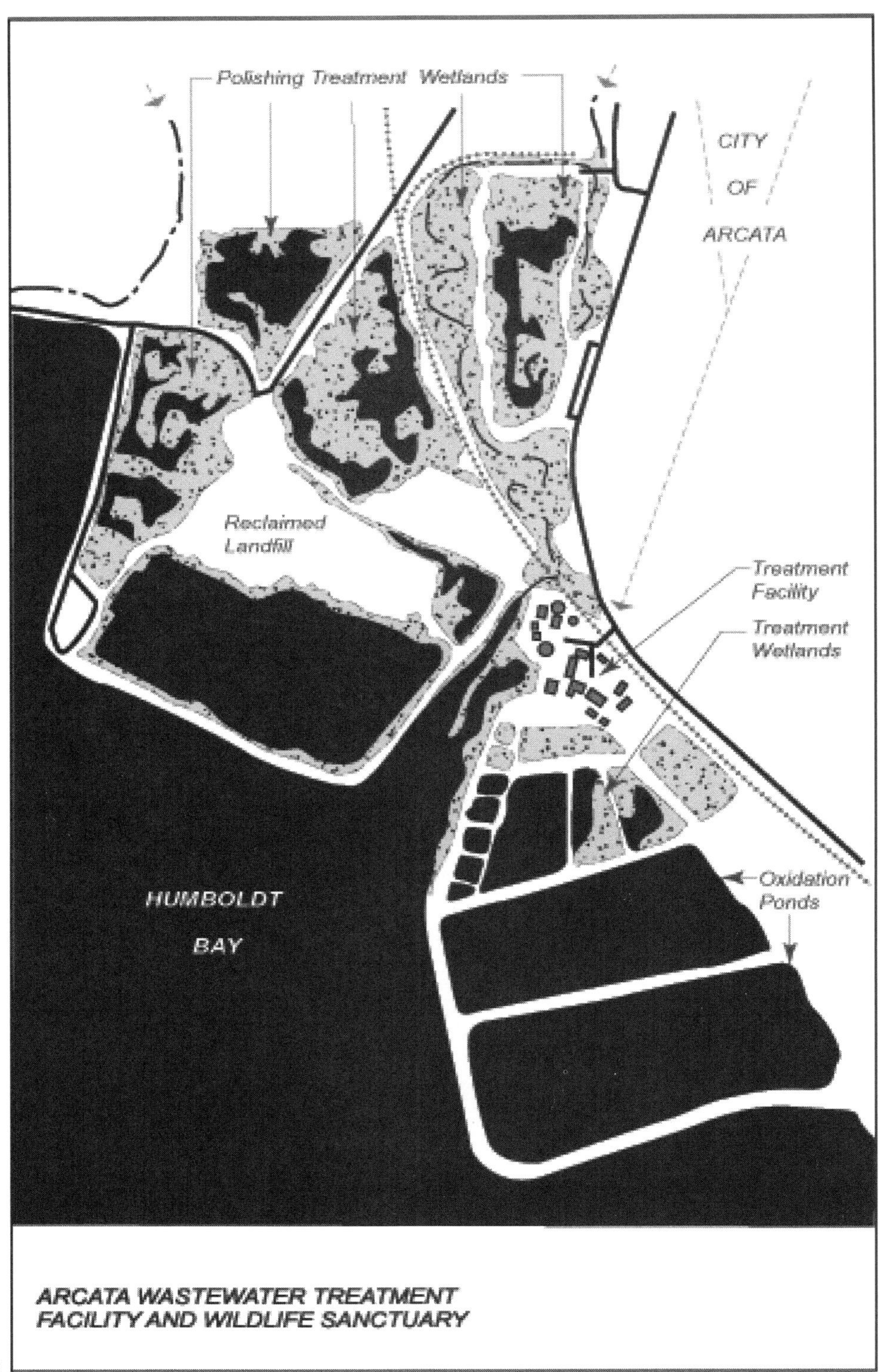

ARCATA WASTEWATER TREATMENT FACILITY AND WILDLIFE SANCTUARY

3-133

3-134

The primary purpose of the 40-acre (16-ha.) intermediate marsh system is to remove suspended solids prior to chlorination and dechlorination. The intermediate marsh system contains 45-ft. (14-m.) stretches of open water spanning the complete width of each of the two cells. These open spaces were designed to provide habitat for fish, which in turn, control mosquitoes. Use of hardstem bulrush in the alternating planted strips also facilitates fish access along open water channels and consequent mosquito predation.

The project cost about half a million dollars (including planning, environmental assessments, and land acquisition), and has achieved good to moderate success in contaminant removal. Further, the new marsh

3-135

system is used for scenic enjoyment and educational study, and has become a major birding area.

REFERENCES

A Guide to Birding In and Around Arcata. Arcata, CA: City of Arcata, 1995.

Bulger, S. *The Beginners' Guide to Birding at the Arcata Marsh*. Arcata, CA: Friends of the Arcata Marsh, 1996.

Environmental Protection Agency. "A Natural System for Wastewater Reclamation and Resource Enhancement: Arcata, California." In *Constructed Wetlands for Wastewater Treatment and Wildlife Habitat: 17 Case Studies*. Washington, DC: Government Printing Office, 1993. 55–66.

Gearheart, R.A "The Arcata Wetlands and Landfill." In *Brown Fields and Gray Waters: Restoring Post-industrial and Degraded Landscapes*, edited by N. Kirkwood and R. France. In prep., 2002.

MacDonald, L. "Water Pollution Solution: Build a Marsh." *American Forests* 100 (1994): 26–29.

The Fort Whyte Centre

(Winnipeg, Manitoba, Canada, Ducks Unlimited)

- nature education
- waterfowl habitat creation
- brownfield reclamation

A vibrant wildlife education center located on a reclaimed quarry and surrounded by agricultural land on the outskirts of Winnipeg contains over 200 acres (80 ha.) of lakes, wetlands, woodlands, and meadows. The lakes, which had originally been excavated to mine clay for cement production, gradually filled in with rain and melted snow to depths of up to 30 ft. (10 m.) and began to attract migrating waterfowl. Converting several lakes into functional wetlands has increased the abundance and diversity of bird life, the motivation for creating the nature center and education program.

Today, over 4 mi. (6 km.) of trails link many of the water features of the site. The large lakes still serve as the major staging areas for migrating waterfowl, which then make ample use of the many smaller wetlands. One small wetland, created by berming off a section of a lake, contains an elaborate network of floating boardwalks designed to allow visitors to experience the wetland more intimately. Nearby, a series of waterfowl gardens display the regional diversity of wetland types, from prairie potholes to boreal bogs and coastal tundra.

Bison
prairie
trout
lake
lake
devonian
Interpretive
Wetlands
lake 3
muir lake
lake 2
FORT WHYTE CENTRE

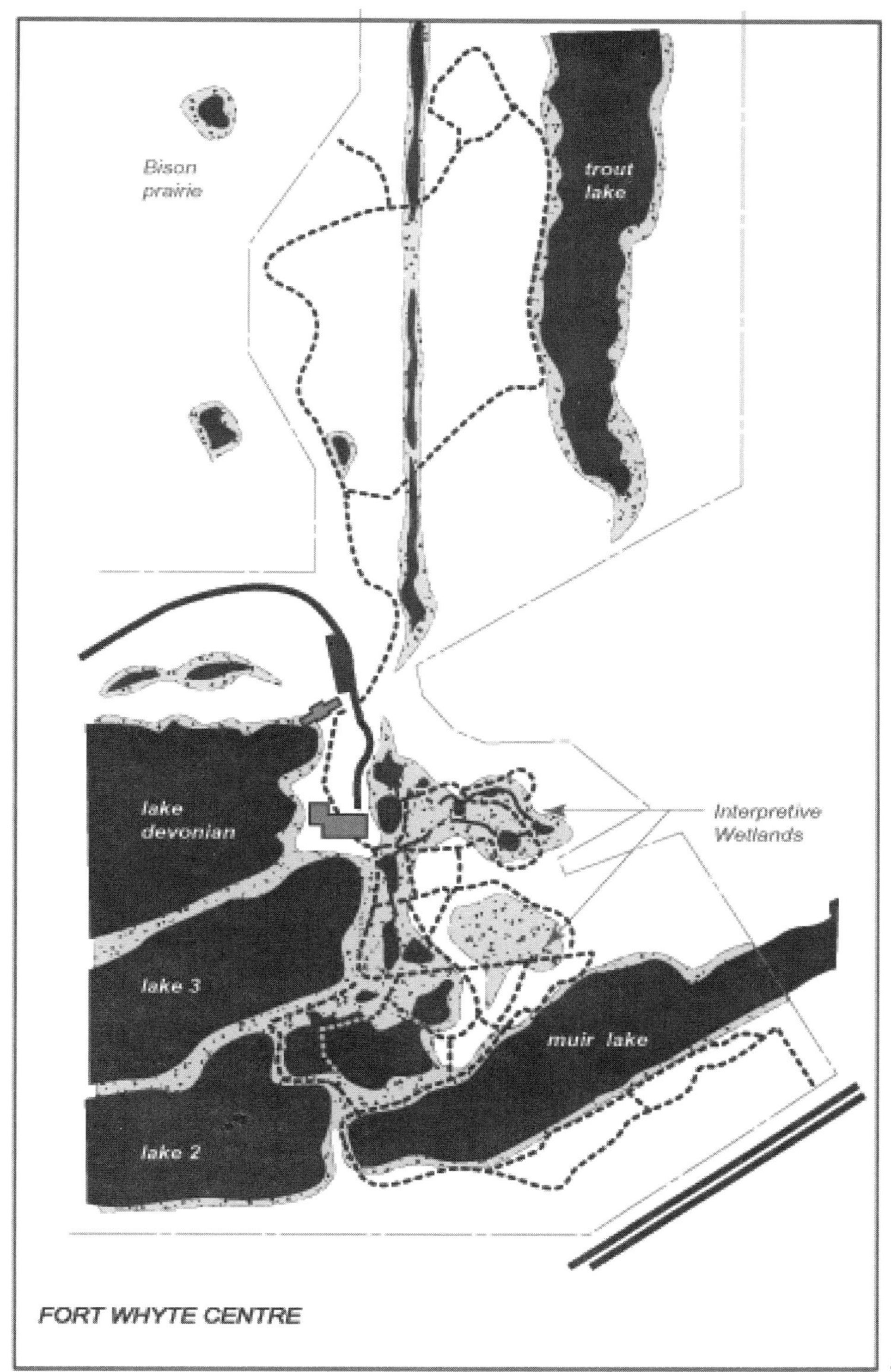

3-136

3-137

Of the 100,000 annual visitors, a third to a half are school children who are served by a large volunteer and professional staff and a diverse education program focusing on the importance of wetlands in waterfowl ecology and management. Recreational activities, such as sailing lessons and sport fishing, are offered on several of the large lakes.

Water is circulated into the created wetlands from the lakes, and a closely monitored system of stop dams between water bodies controls seasonal hydrologic variability. In the near future, parking lot runoff as well as toilet and restaurant gray water will be directed into a constructed

3-138

wetland before discharge into one of the lakes. The functional utility of treatment wetlands for water purification will be featured in an expanded education program concerning sustainability.

Other site elements include agricultural demonstration plots, alternative composting systems, a 70-acre prairie with the largest urban herd of plains bison in North America, and a large interpretive center containing historical landscape dioramas, a gift shop, an indoor waterfowl wintering room, lecture theaters, and an aquarium featuring local aquatic life.

Plans are underway to construct a floating fieldstation over a bay on one of the lakes to allow students to engage in water monitoring and climate studies. Finally, paying attention to the larger landscape, a green corridor connects the Fort Whyte Centre through an expansive nearby urban forest to a major park located along the banks of one of the city's two main rivers. This corridor provides wetland access to terrestrial mammals.

REFERENCES

Caldwell, L. "Wetland Park and Industrial Site in Winnipeg." In *Brown Fields and Gray Waters: Restoring Post-Industrial and Degraded Landscapes.*

Fort Whyte Centre. "Reaching New Horizons." Brochure, 1999.

Scarth, A. "The Wildlife Foundation of Manitoba: Coming of Age." *Branta Newsletter*. 6 (1988): 1–4.

The Oregon Garden

(Silverton, Oregon, U.S.A., Mayer/Reed–Interfleuve Inc.)

- wastewater treatment
- mitigation bank credits for wildlife habitat creation
- integration into overall garden to showcase local wetland plants
- source of reclaimed water for garden maintenance

This project ably demonstrates the many ways to combine functions in innovative wetland creation. The city of Silverton had two goals to accomplish: first, they needed to treat 500 gal. (2,000 l.) per minute of wastewater effluent; and second, they had to mitigate for 7 acres (3 ha.) of wetlands that were unavoidably lost due to industrial development. At the same time, planning was underway for "The Oregon Garden," a world-class showcase of plant diversity that would offer research and educational opportunities and conference facilities.

The result of these goals took the form of three created wetlands for final wastewater polishing through biotreatment, receipt of mitigation credits for wildlife habitat creation, and an exciting and aesthetically

3-139

3-140

appealing feature to serve as the central tourist attraction of the new garden complex.

In the A-Mazing Water Garden, serving as the major entrance feature to the overall 240-acre (96 ha.) Oregon Garden, the topographic challenges of the site were met by constructing a series of sixteen terraced, rice paddy-like wetland cells covering 5 acres (2 ha.). Through these terraces water falls 37 ft. (11 m.) over the 1000-ft. (300-m.) length. It is here that the city's wastewater treatment plant discharges its highly nutrient-laden effluent during the spring to autumn period. The wetland has been designed to accommodate a doubling of effluent discharge over the next decade.

The A-Mazing Water Garden also functions as wildlife habitat, displays botanical features, and provides numerous opportunities for education and research. Access to the upper wetland is restricted to several remote observation knolls, built of material excavated during wetland construction and designed to provide minimal disturbance to wildlife. A 40-ft. (12-m.) wide buffer of upland trees is provided to further separate the wildlife wetland from adjacent garden activities. An outreach program is planned which will focus on educating homeowners and school children about wetland ecology and management.

The bottom of the wetland is designed as an ornamental display pond in which water lilies, lotuses and other aquatic plants of a variety of shapes and sizes provide a textured landscape. A pedestrian bridge serves as the divide between the natural wetland terraces upstream and the refined wetland gardens in the pool area below. There, a network of curvilinear pathways snake through ribbons of emergent vegetation and

3-141

ornamental grasses to create "a virtual maze of fun and disorientation," as the designer stated (hence, the "a-mazing" name of the wetland).

Water then flows into a 1.2-acre (0.5-ha.) wildlife wetland containing native trees, shrubs, and emergents for further loss-replacement mitigation credits. Finally, this water moves into the last wetland, also designed as a series of cells and using native plants for more mitigation credits. This 10-acre (4-ha.) wetland was kept unlined to promote infiltration and groundwater recharge.

As a further precaution against environmental damage, water from the entire wetland complex is discharged into nearby Silver Creek only when river flows are sufficiently high. During the dry season, all discharge from the wetlands is recycled onsite for garden irrigation.

REFERENCES

Fields, K.J. "A-mazing Effort: Wetlands, Wildlife and Resource Conservation Flourish in Water Garden." *DJC Magazine* (August 2000): 93–94.

Koonce, G. "Concepts and Technology of the A-mazing Water Garden: A First Phase Project of The Oregon Garden." *Interfleuve Internal Document*, 1998.

Mayer-Reed, C. "Concepts and Technology of the A-mazing Water Garden: A First Phase Project of The Oregon Garden." *Technical Innovations in Landscape Architecture 6* (1998): 1–3.

Underwood, T. "The A-mazing Water Garden." *Land Forum* 06/108 (1999): 35–38.

The Wetland Centre

(London, England, Wildfowl & Wetlands Trust)

- wildlife habitat restoration
- wetland ecotourism and conservation education

Located halfway between Westminster and Kew Gardens in London is one of Europe's most impressive ecological restoration projects, and the largest area of created wetland habitat in any capital city throughout the world. The legacy of noted waterfowl conservationist and painter Sir Peter Scott, the project has on its board such celebrities as naturalist Sir David Attenborough and Charles, the Prince of Wales.

The scale of the project is immense, covering over a hundred acres reclaimed from four abandoned Victorian reservoirs situated on the banks of the Thames. Sale of a portion of the land to a developer for an upscale housing project raised enough money to initiate the restoration of the rest of the site. In the end, 600,000 cu. ft. (500,000 cu. m.) of soil were relocated and sculpted into over thirty reformed lakes, ponds and marshes, regulated by twenty-seven water control sluices, and containing 150 ft. (46 m.) of boardwalk, twenty-seven bridges, 2 mi. (3 km.) of pathways, seven waterfowl observation hides, and 27,000 trees and 300,000 aquatic plants. The large visitor center offers a two-level observatory, audiovisual lecture theater, education center, restaurant, shop, and art gallery. The design team from the WWT (Wildfowl & Wetlands Trust) included hydrologists, civil engineers, soil scientists, landscape architects, and architects, in addition to ornithologists.

Water is pumped to the site from a non-tidal upstream location on the

3-142

residential area
residential area
Observation Hide
world wetlands
Observation Tower
Wader Scrape
THAMES RIVER
Visitor Center
waterlife wetlands
WETLANDS CENTRE

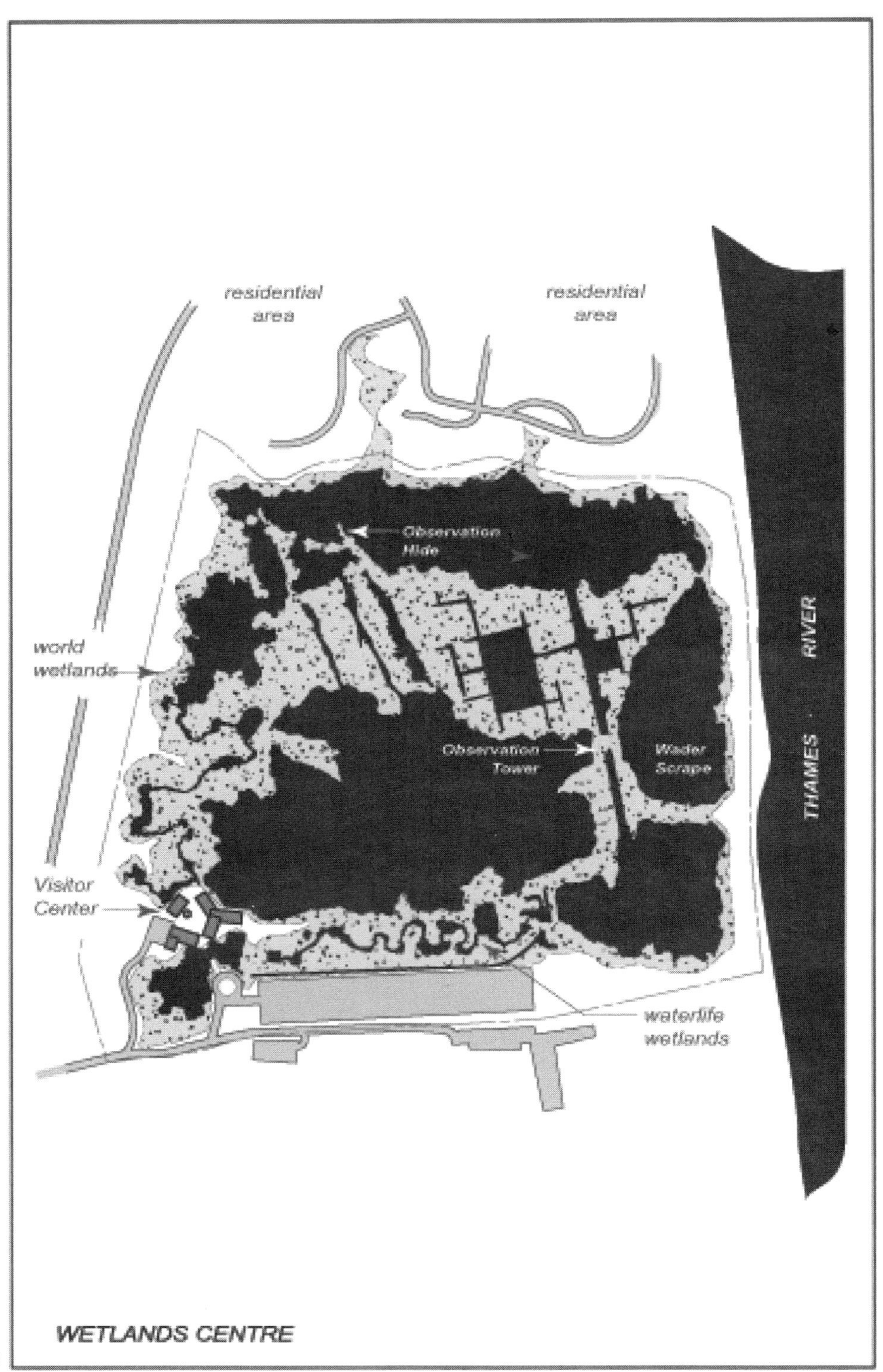

3-143

Thames, and after transit through the wetlands, where each water body is precisely and independently regulated, it is discharged to the river. Because no new material (e.g., soil) could be brought onto the site and no old material (e.g., reservoir concrete liners) leave it, much attention was devoted to careful site work. For example, the entire restored landscape is sealed with clay to maintain its hydrological isolation. Islands were oriented to reduce wave energy and constructed to encourage waterfowl use.

The majority of the site is a natural area devoted to fostering feeding, breeding, and roosting of waterfowl in a variety of aquatic habitats including open water lagoons, lakes, grasslands, and mudflats. Two smaller areas are designed as outdoor instructional exhibits. One functions as a Wetlands of the World zoo in which thirteen international wetland habitats (subarctic boreal bogs, tropical swamps, etc.) are recreated and where clipped-wing birds from each region reside. The other area highlights the human use of wetlands in the United Kingdom, such as for food production, thatch farming, flood control, and sustainable gardens.

A large outreach program educates people about the role of wetlands in promoting biodiversity and sustainability, and an active research program focuses on understanding the effects of habitat creation on waterfowl.

REFERENCES

France, R. "Barnes Reservoirs to the Wetland Center (London)." In *Reclaimed!: Recovery Processes and Design Practices for Post-Industrial Landscapes.*

The Wetland Centre. Brochure, 2000.

"Wetland Centre Monitoring Project: 20th Progress Report: 1st July–30th September 2000." In *The Wetland Centre Report*. London, UK: 2000.

3-144

Hangzhou Botanical Nursery

(West Lake, Hangzhou, China, City of Hangzhou)

- aesthetic water garden
- minimal water quality improvement

The Hangzhou Botanical Nursery was built in 1956 and occupies sixty-three acres near the shore of the famed West Lake. This region receives over twelve million tourists a year, drawn by the scenic beauty that has inspired poets for over a millennium. The primary purpose of the garden is to highlight its considerable collection of native trees and bonsai plants. Because the art of water gardening has a long established tradition in China, and there are already many wonderful water features in and around West Lake itself, the nursery decided to create its own wetland to frame its central building.

Many of the elements of traditional water gardens are present here, including complex shape geometry, sculpted banksides, islands, meditation benches, observation and fishing platforms, zigzag walkways, and pagodas. Each structure is placed to frame particular views of the others. The central building lies along one entire side of the wetland, extending over the water and offering spectacular views from the restaurant within. Stairs to the flat-topped roof provide access for overviews of the entire wetland.

Presently, however, this artificial wetland functions primarily as an aesthetic garden, as it is coupled to the watershed only during extreme storm events when a polluted river overflows into the nursery. In the watershed case studies described in Chapter Two of this book, a plan by

3-145

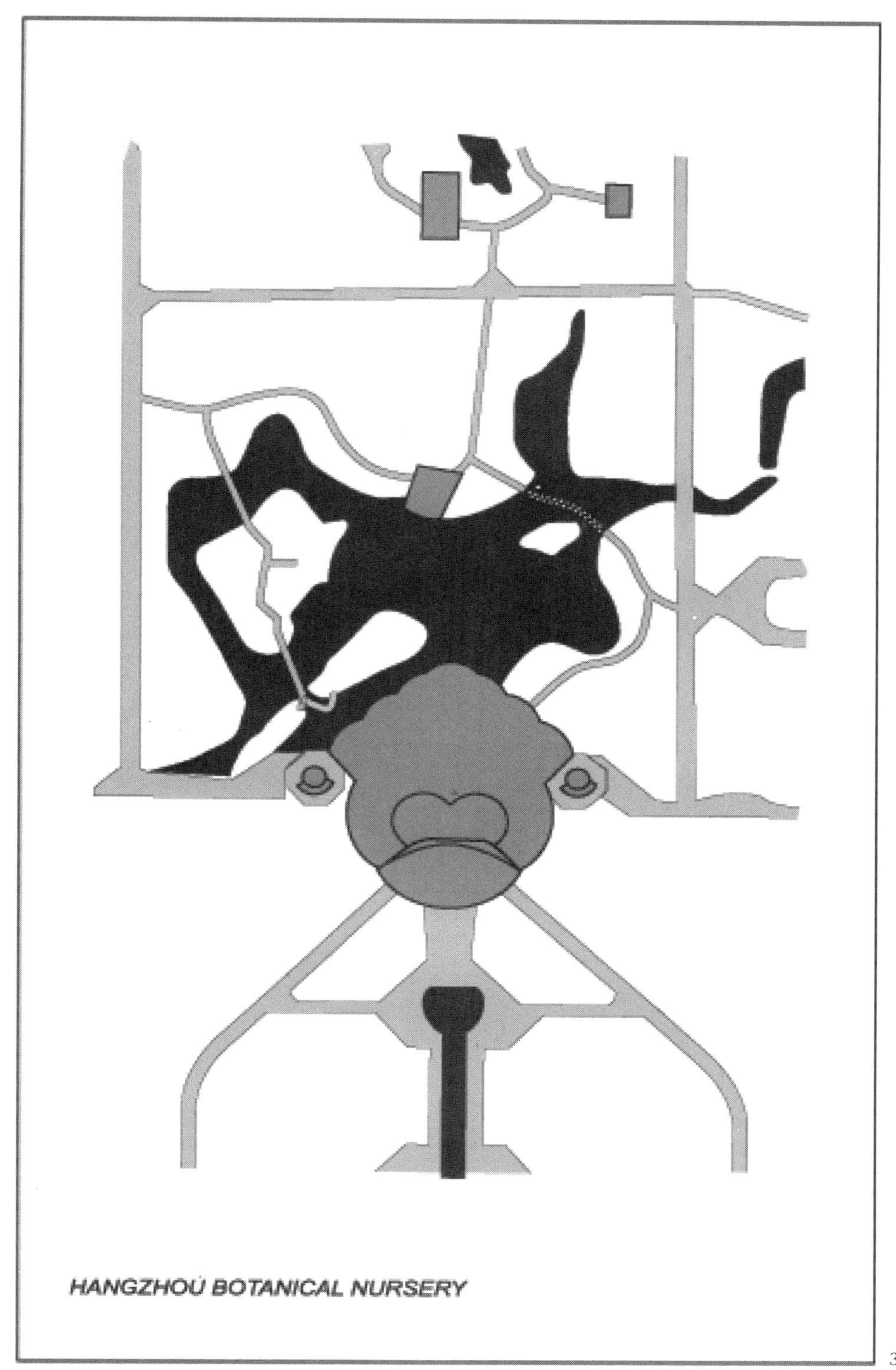

3-146

3-147

students from the Harvard Design School shows how the present scenic wetland can be easily converted into a truly functional treatment wetland used for final polishing of the river's nutrient load before its discharge into West Lake.

REFERENCE

Gang, C. *West Lake Poetics.* Hangzhou, China: Zhejiang Photographic Press, 1996.

Meadowbrook Pond and Wildlife Habitat Park

(Seattle, Washington, U.S.A., Lydia Alredge, Peggy Gaynor, and Kate Wade–Seattle Public Utilities–Seattle Arts Commission)

- stormwater detention
- habitat restoration
- public art display
- education

Thornton Creek is Seattle's largest watershed, and like most urban watersheds it experiences stormwater flooding and erosion scouring. As part of a project in watershed restoration, including stream daylighting (excavating buried waters) and salmon habitat reconstruction, a former sewage treatment plant was converted into a 9-acre (4-ha.) riparian wetland.

A complicated system of inflow weirs, overflow diversions, and creek realignments were left visually prominent in this design to instruct visitors about the effects of urbanization on water management. Interpretative signs, in addition to discussing this issue, instruct the reader on the importance of habitat restoration for riparian and wetland species in urban settings. One example:

> ***How does this place work?***
> This site functions as an art piece—a refuge from its urban surroundings—as it fulfills its stormwater management role. When stream volume reaches a certain height, water flows into the pond where it slows down, dropping sediment at the rock weir. Water then continues into the pond via the main channel, flowing north under the bridge as well as by way of a side channel on the south side of the bridge. Water from the pond rejoins the stream. At extreme volumes, the stream and pond will back up and overflow into an outfall pipe which bypasses lower Thornton Creek altogether, runs underground to the east, and daylights at Matthews Beach on the shore of Lake Washington. This constructed system helps prevent flooding, which was typical in the watershed, and improves water quality by allowing fast-moving turbid water in the stream to slow down and drop sediment where it can be removed from the system.

Native plantings of both wetland emergents and riparian trees attract numerous waterfowl, as does the presence of islands, several of which retain their trees. A nearby school uses the site to study wetland horticulture.

Great attention was paid to creating this public amenity, which has become widely used. Bridges, trails, and boardwalks allow visitors to

inflow
street
Weir
Green-
houses
street
street
outflow
Footbridge and
Stormwater Pipe
MEADOWBROOK POND

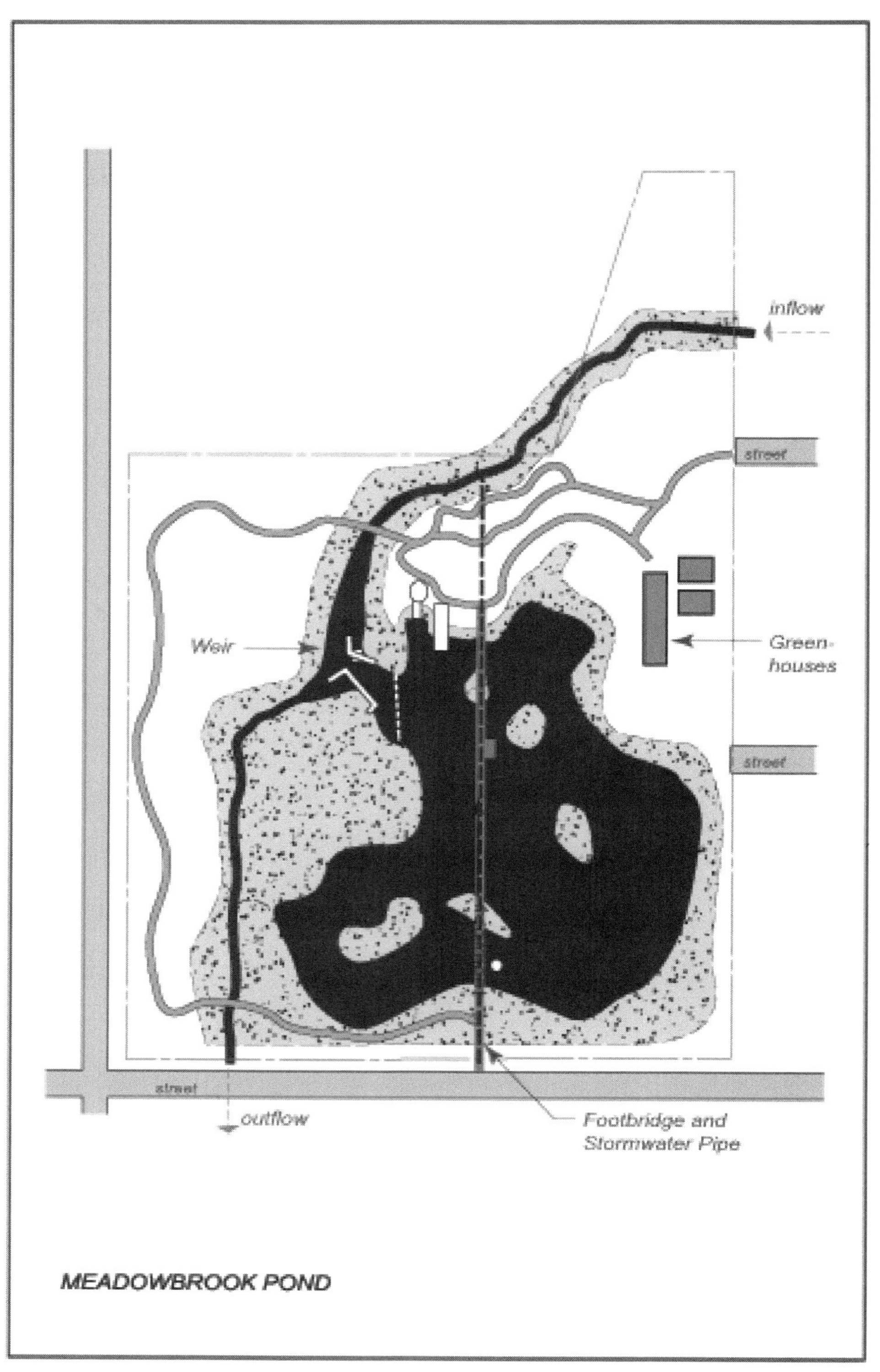

3-148

3-149

meander throughout the site. Excavated material from the detention pond was used to sculpt landforms around which the pathways wind. Several art installations are featured in a plaza with a seating area, including a reflecting sound mirror that collects and focuses the noise of water as it rushes over one of the outlet weirs.

REFERENCE
MacElroy, W.P., and D. Winterbottom. "Stormwater Ponds." *Landscape Architecture* 4/00 (2000): 48–54.

3-150

Water Pollution Control Laboratory

(Portland, Oregon, U.S.A., Murase Associates)

- stormwater treatment
- education
- aesthetics

The research laboratory of the Portland Bureau of Environmental Services, whose task is to monitor urban water quality, was conscious of the presence of nearby signs containing health warnings about the discharge of combined sewer outfalls after storm events. The agency wanted to lead by example and create a demonstration project in which stormwater treatment, rather than being hidden, is dramatically made visible.

The site, located on the banks of the Willamette River, drains a 50-acre (20-ha.) residential and commercial neighborhood which includes at least one industrial storage compound and had previously contained a derelict factory. As a result of a lengthy community participation process in which the neighbors voiced their concerns, the concept of a traditional detention basin was changed to that of a vibrant wetland which celebrates stormwater as few other projects ever have.

Innovation begins surrounding the building itself. The parking lot contains experimental drainage swales in which electronic augmented weirs feed data directly into the laboratory for constant monitoring. Further, the building dispenses with traditional gutters and instead features extended roof sputters which feed rain gardens near the entrance.

It is the wetland itself, however, which is designed to draw most of the attention, for here stormwater management is elevated to an art form. Water from neighborhood pipes emerges into the wetland through a beautiful, rock-lined surge dissipater flume. After depositing particulates, water is discharged laterally through perforated weepholes in the side of the raised plume bed. The upper section of the open pond contains a semicircular basalt wall rising from 2–8 ft. (0.5–2 m.) above the water. By deflecting water flow and thereby increasing residence time, the wall improves water quality. During extreme storm events, water is directed to an overflow pipe feeding directly into the river; otherwise, stormwater infiltrates the ground for further purification before seeping into the river.

Great attention was paid to vegetation, ranging from a littoral fringe of traditional wetland emergents such as cattails and soft rush to a surrounding riparian garden which includes ornamental grasses. Although it was not intended to provide habitat, waterfowl and fish have become residents. Finally, the view from the office windows overlooking the wetland is superb, no doubt decreasing stress and increasing morale.

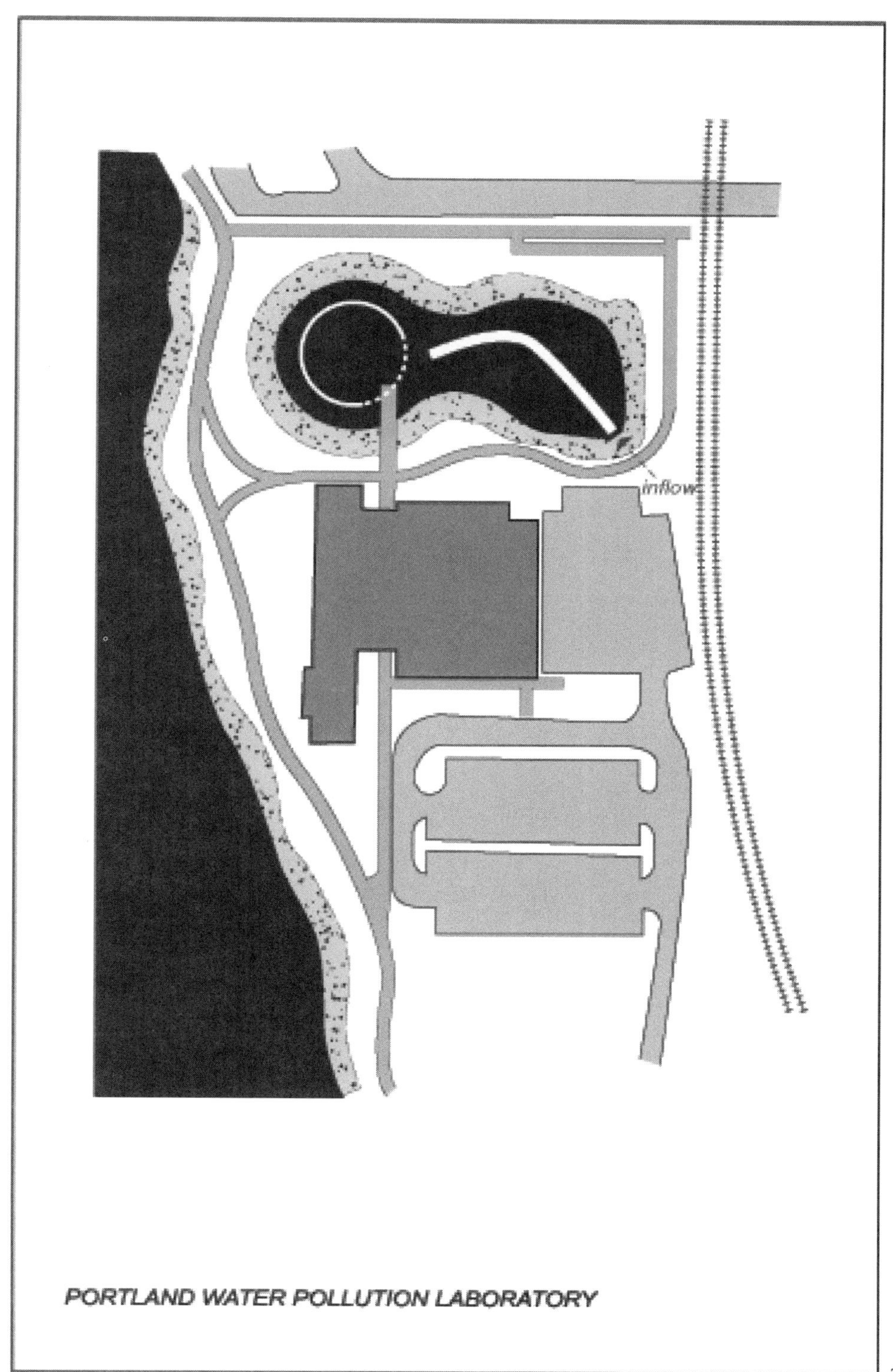

3-151

3-152

3-153

REFERENCES

Thompson, J. W. "The Poetics of Stormwater." *Landscape Architecture* 1/99 (1999): 58–63.

Liptan, T., and R. Murase. "Stormwater Gardens as Urban Infrastructure." In *Handbook for Water Sensitive Planning and Design*.

Potawot Health Village

(Arcata, California, U.S.A., Humboldt Water Resources)

- creation of human amenities
- wetland restoration
- stormwater retention

A health facility is currently under construction in Arcata, California, that is designed to serve nine tribes with more than 13,000 Native Americans. The working mandate of the United Indian Health Services is to create a 40-acre (15-ha.) site surrounding its clinic that will integrate health, community, and environment into a multiple-use landscape. Here, in addition to a sweathouse, dance pit, and wellness garden, wetlands will be used to address spiritual and cultural aspects of traditional healing.

Stormwater will be managed onsite through reduction of impervious surfaces, use of vegetated drainage swales and sediment traps, and creation of a system of interlinked wetlands. The latter are carefully designed to fit into the topography in such a way as to reduce the threat of flooding to new developments planned for the site.

The existing ephemeral wetlands have become degraded due to changes in hydrology in the surrounding landscape. These will be restored through excavation and reestablishment of native grass meadows and woodlands as part of a concerted effort at wildlife enhancement.

Above all, these restored wetlands are being designed to promote concepts of human rest and recovery, the cornerstone of the center's approach to health and healing. Recreation will be encouraged along a

3-154

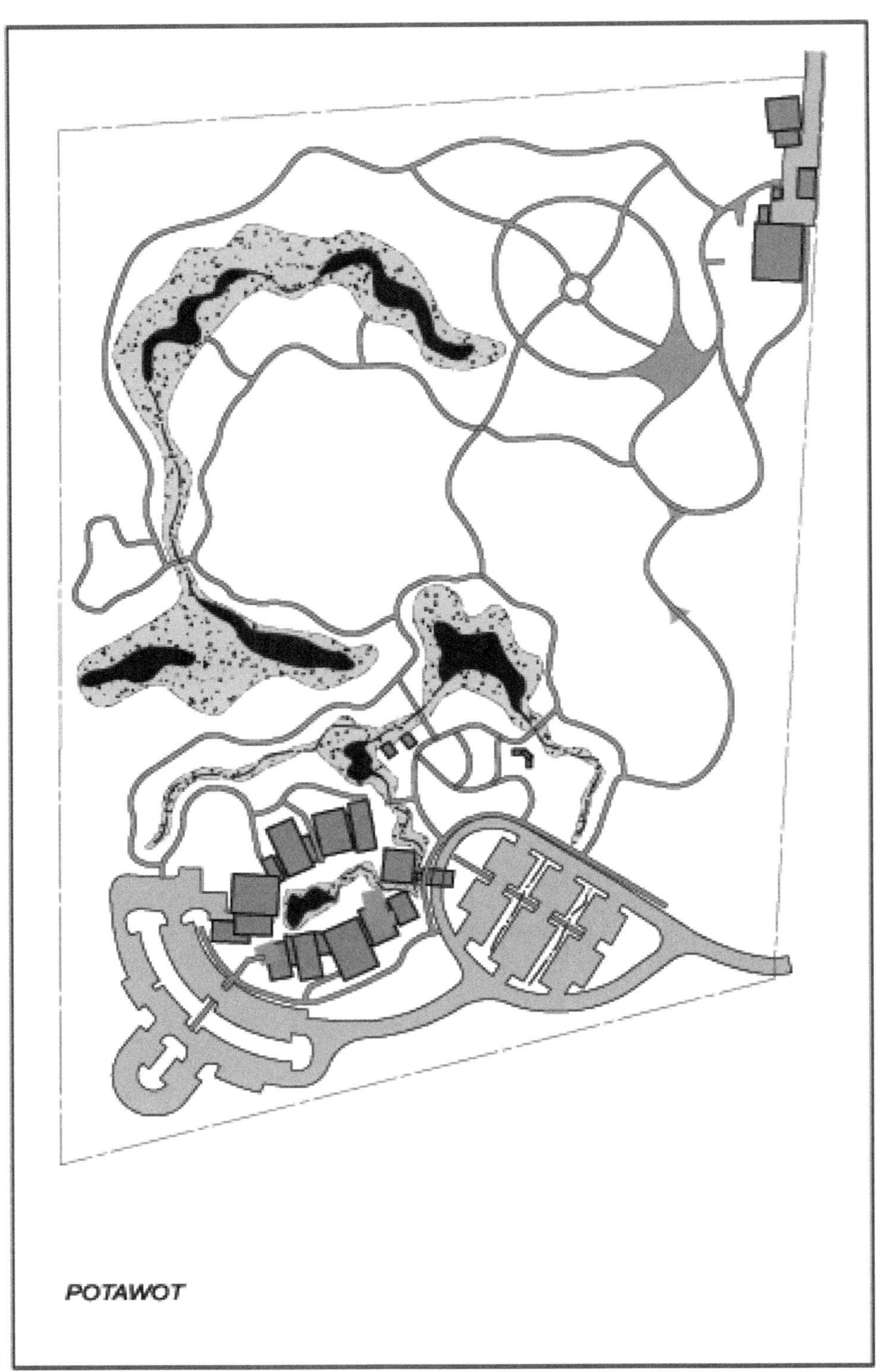

3-155

3-156

network of trails and benches designed for walking, jogging, wildlife observation, and picnicking, in addition to meditation and spiritual reflection.

Plants were carefully selected. Certain wetland areas will be devoted to the production of native plants to be used for cultural purposes such as crafts and traditional medicine.

Visitors will learn about wetland functions of stormwater storage, water quality improvement, wildlife habitat restoration, and native plant healing through interpretive signage and kiosks.

REFERENCES

Kadlecik, L. "United Indian Health Services' Potawot Health Village: Integrating Health, Community and the Environment." Unpublished. Arcata, CA: 2000.

Kadlecik, L. "Wetlands and Wellness." In *Brown Fields and Gray Waters: Restoring Post-Industrial and Degraded Landscapes.*

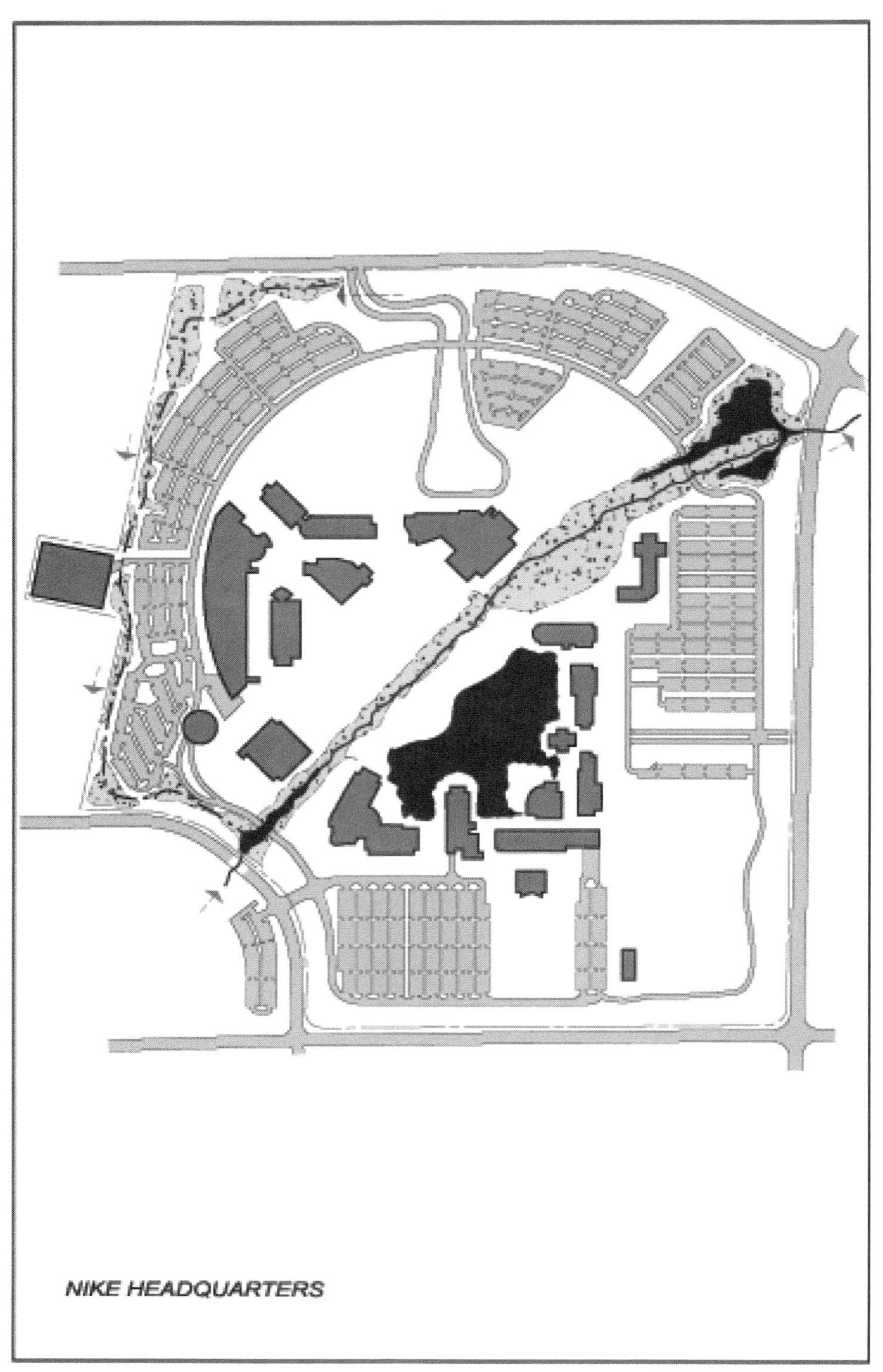

3-157

3-158

Nike Headquarters

(Beaverton, Oregon, U.S.A., Mayer/Reed–Murase Associates)

- stormwater retention
- water quality improvement
- habitat restoration
- recreation and employee wellbeing
- creation of site identity

Nike, in a move to instill employee confidence after a spate of bad press, and as a move toward becoming more environmentally friendly, has concentrated its formerly scattered offices into a single campus. The enlarged campus, situated within an encircling 20-ft. (7-m.) berm, is designed as a sanctuary and refuge for employees. The design contrasts internal paved plazas and manicured landscapes with surrounding wilderness of which wetlands form a major part.

A spectacular curving stone wall with a self-contained seeping waterfall crowns the north entrance. The impression given is that this water feature is hydrologically connected to the adjacent wetland, the first of ten that wrap around one side of the site. As a positive gesture to the neighborhood, stormwater enters the site and moves through a complicated series of created wetlands, each being regulated by weirs. These wetlands are also used to manage all onsite runoff from the fringe parking lots. After processing through the wetland complex, water is transported to the central stream that bisects the campus. Here, streambanks

3-159

have been excavated in several locations to restore riparian wetlands. These wetlands, located within the existing native woodland, now provide havens for numerous waterfowl as well as increasing flood storage capacity.

Because the staff are very active, everything was done to encourage them to get outside during their breaks. A perimeter jogging track that winds its way alongside and between the stormwater wetlands is so popular that employees even return to use it on weekends!

REFERENCE

Bennett, P. "Worlds Apart." *Landscape Architecture* 8/00 (2000): 60–69.

Montreal Water Park

(Montreal, Quebec, Canada, City of Montreal–Environment Canada)

- wastewater treatment
- river water purification
- tourism and education
- habitat recreation

Ile Sainte-Helene, located in the middle of the St. Lawrence River across from downtown Montreal, used to house fortifications built by the British to defend the city from American attack. The location, however, is better known as being—along with the adjacent Ile Notre Dame, created by dumped excavation materials—the location of the 1967 Expo. The Island

Wastewater
treatment wetland
BioDome
ST. LAWRENCE
RIVER
River purification
wetland system
MONTREAL WATER PARK

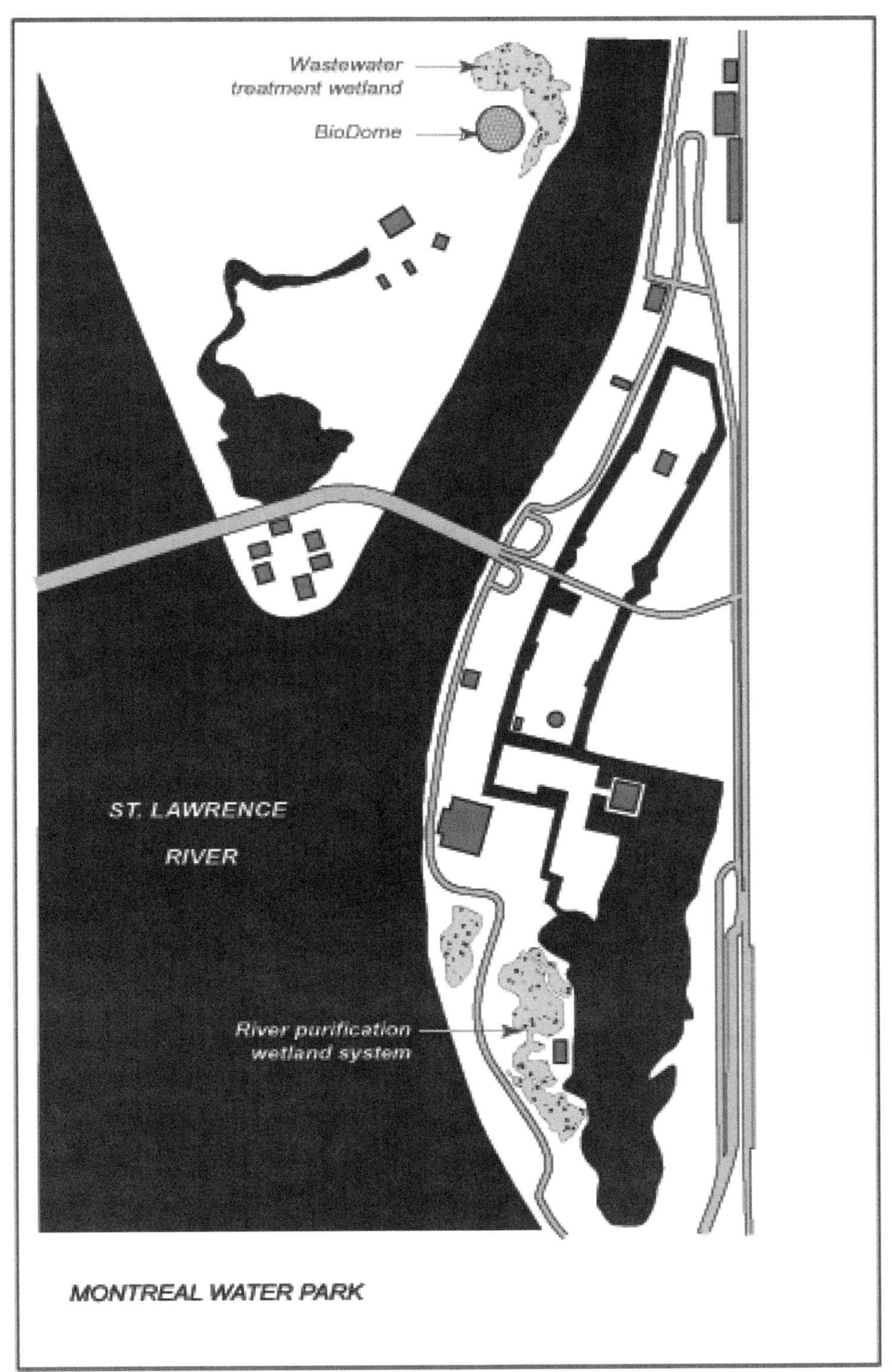

3-160

3-161

Trilogy complex, occupying the former World's Fair park, is designed to raise public awareness about the environmental importance of wetlands.

At the old United States pavilion, which has been converted into "The Biosphere," an environmental education center, a pilot project demonstrates to the public and industry that wetland plants can provide simple, economical and effective treatment of wastewater for small temperate communities. After preliminary treatment in a septic tank, building wastewater enters two parallel subsurface-flow beds filled with common

3-162

reed grass at a rate of 4000 gal. (15,000 l.) per day. Water then flows into two successive surface wetlands containing cattails, in addition to several other species selected for their contaminate uptake and water oxygenation abilities. The total treatment surface area is 8600 sq. ft. (800 sq. m.), which removes about 80 percent of the suspended solids, oxygen depleting substances, and phosphorus.

Located beside the old Canadian pavilion and the new Grand Prix race track is a half-acre wetland created to educate visitors about the resource industry of peat farming in northern Quebec. Over a thousand peat blocks, each weighing about 1200 lbs., were cut out of a bog from James Bay, transported 900 mi. (1,900 km.) south, and reassembled like a giant puzzle. Special attention to surface and subsurface hydrology has maintained the incongruous wetland in a semi-natural state.

Faced with the challenge of developing a small lake on the island for such recreational activities as sailing and swimming, a series of wetlands totaling 50 acres (20 ha.) were created to filter and treat water pumped from the surrounding river. Once the river water fills the lake, it is continually recirculated through four wetlands and then UV-treated before being discharged back into the lake over a small set of rapids. The fact that thousands of visitors are immersing themselves in water that has been purified by passing through the wetlands is an incredible opportunity for education. Consequently, there is an interpretative kiosk describing the roles of wetlands in water recreation.

REFERENCES

City of Montreal and Environment Canada. "Island Trilogy Sustainable Development: From Ideas to Action: Parc Jean-Drapeau and Biosphere." Brochure. Montreal, CA: City of Montreal, 2000.

Environment Canada. "St. Lawrence Technologies: An Extended Wastewater Treatment System." Brochure. Montreal, CA: 1996.

Environment Canada. "The Biosphere: Where Water Tells Tales." Brochure. Montreal, CA: Ecowatch Centre, 2000.

Vincent, G. "Artificial Marshes to Maintain Water Quality: The Beach of Ile Notre-Dame." Water Pollution. *Research Journal of Canada* 27 (1992): 327–339.

the tunnel
the grotto
the knoll
the passage
the release
WATERWORKS GARDEN

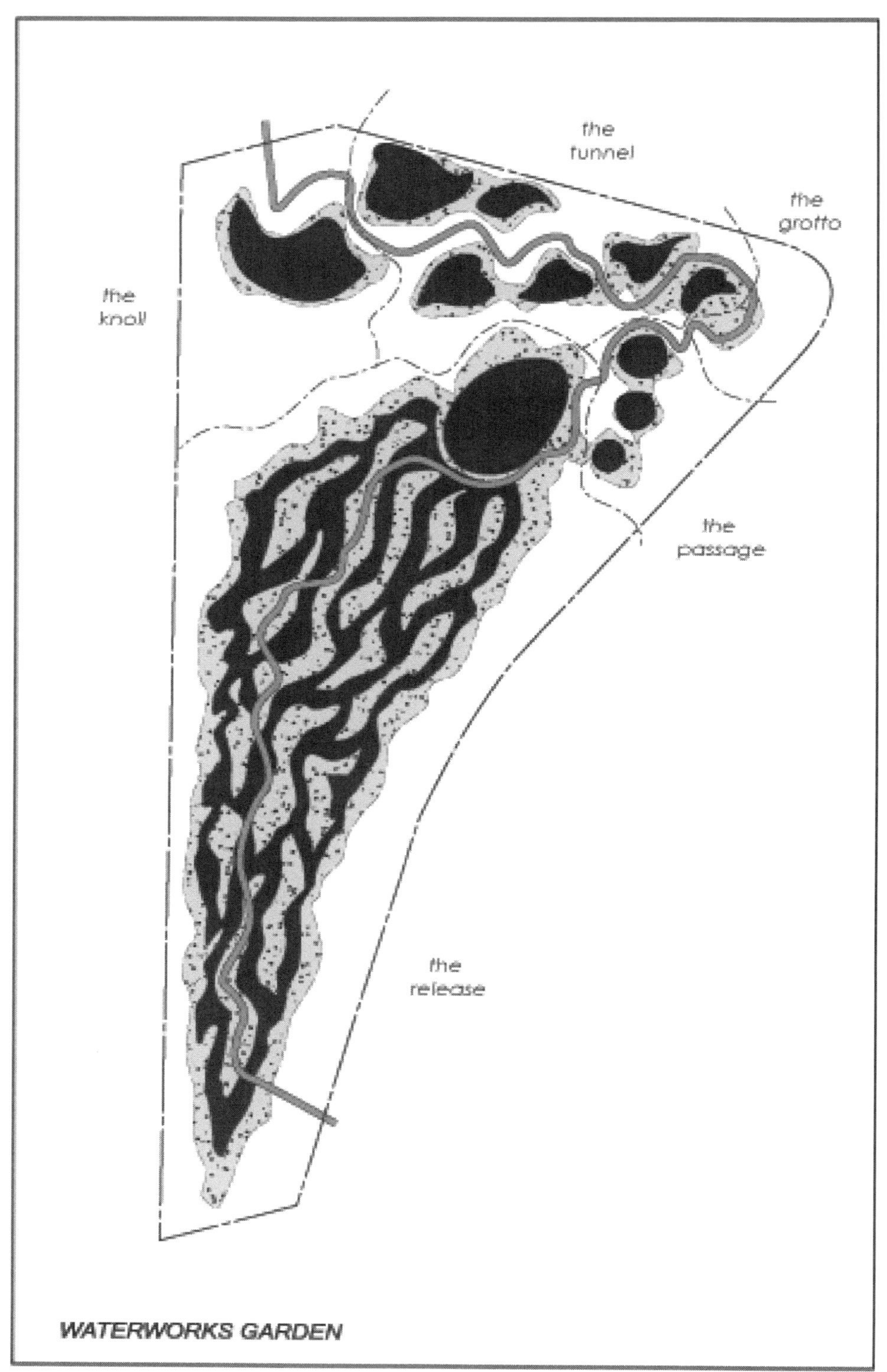

3-163

Waterworks Garden

(Renton, Washington, U.S.A., Lorna Jordan–Jones and Jones)

- stormwater treatment
- public amenity
- education and art activism

Following a long consultation process, necessary to convince civil engineers about development feasibility, a public art project consisting of 8 acres (3 ha.) of created wetlands and settlement ponds was built to treat runoff from the 50 acres (20 ha.) of roads and parking lots at a sewage treatment plant.

Due to space limitations, it was necessary to construct the wetland treatment park on the site of a former golf course fairway situated uphill from the sewage treatment plant. Over 20,000 cu. yd. (15,000 cu. m.) of earth had to be moved to terrace the wetland into the hillside. From the air, the overall design of the project is meant to resemble a giant flower. There are five programatic "rooms," designed to symbolize a journey from the civilized to the wild, which follow the course of water as it moves downhill.

Stormwater runoff from the sewage plant is pumped up to the "Knoll," a colonnade of basalt columns leading to an overlook. In a deliberate move to make infrastructure visible, visitors can observe the pumped water through a grate between the columns. Water then enters the first of eleven settling ponds, where particulates are precipitated. These ponds are lined with clay and recycled ground glass and are

3-164

3-165

fringed with emergent plants. All water features are hydrologically controlled with weirs, pumps, and overdrains.

Through the "Funnel" region the path snakes its way around the leaf-shaped ponds to enter the "Grotto." The Grotto, constructed from recycled marble and a concrete derivative, is designed as a place of contemplation. Water continues to move through the "Passage," a system of round settling ponds until it reaches the "Release," where the flow paths break up into multiple channels that work their way through the natural-looking (but artificially created) wetland. Finally, cleansed water is discharged into a creek at the bottom of the hill, located beside the sewage plant.

In addition to 9,000 wetland plants, trees native to the Pacific Northwest were planted, some selected for their ethnobotanical importance to Native Americans of the region. These were planted symbolically; for example, trees identified with bad omens were used at the top of garden where the water is still contaminated. Interpretive signage conveys the message about the purification role of wetland plants.

The park is well used by office workers and people from the neighborhood. Joggers often wind their way up the "Water Walk Path" to the Knoll, and back down to the regional bike path located at the bottom of the hillside along the creek.

REFERENCE

Leccese, M. "Cleansing Art." *Landscape Architecture* 97/1 (1997): 70–76.

Zhou Zheng Garden

(Suzhou, China)

- private scenic amenity
- popular tourism
- garden history

In the middle of the 16th century, a retired censor created what is generally regarded as one of the most beautiful gardens in the world. Water covers three-fifths of the 10-acre (3-ha.) "Humble Administrator's Garden." Like all of Suzhou's gardens, this one is based on contrasting bustling urban life with a secluded retreat for meditation. A visitor need not have to leave the city to find peace and stillness.

As in ancient Chinese gardens, the Humble Administrator's Garden lacks unifying symmetry and order, and instead allows the visitor to revel in the process of discovering a series of beautiful spaces. Walls within walls are used to create a feeling of sanctuary, and structures such as verandas, porches, and covered zigzag bridges make no distinction between exterior and interior. An overall austerity makes the experience of discovery more intense.

Islands, linked by numerous bridges, break up the three major water bodies into small, intimate parcels. Strategically arranged moon gates allow framed views of the wetlands, as if in a painting. Terrestrial plants in such scholar's gardens were selected for symbolic meaning, with color and decoration not being important. In the water, lotuses dominate, their willowy blossoms arranged in purposeful contrast to the hard surface of nearby rock sculptures and buildings.

3-166

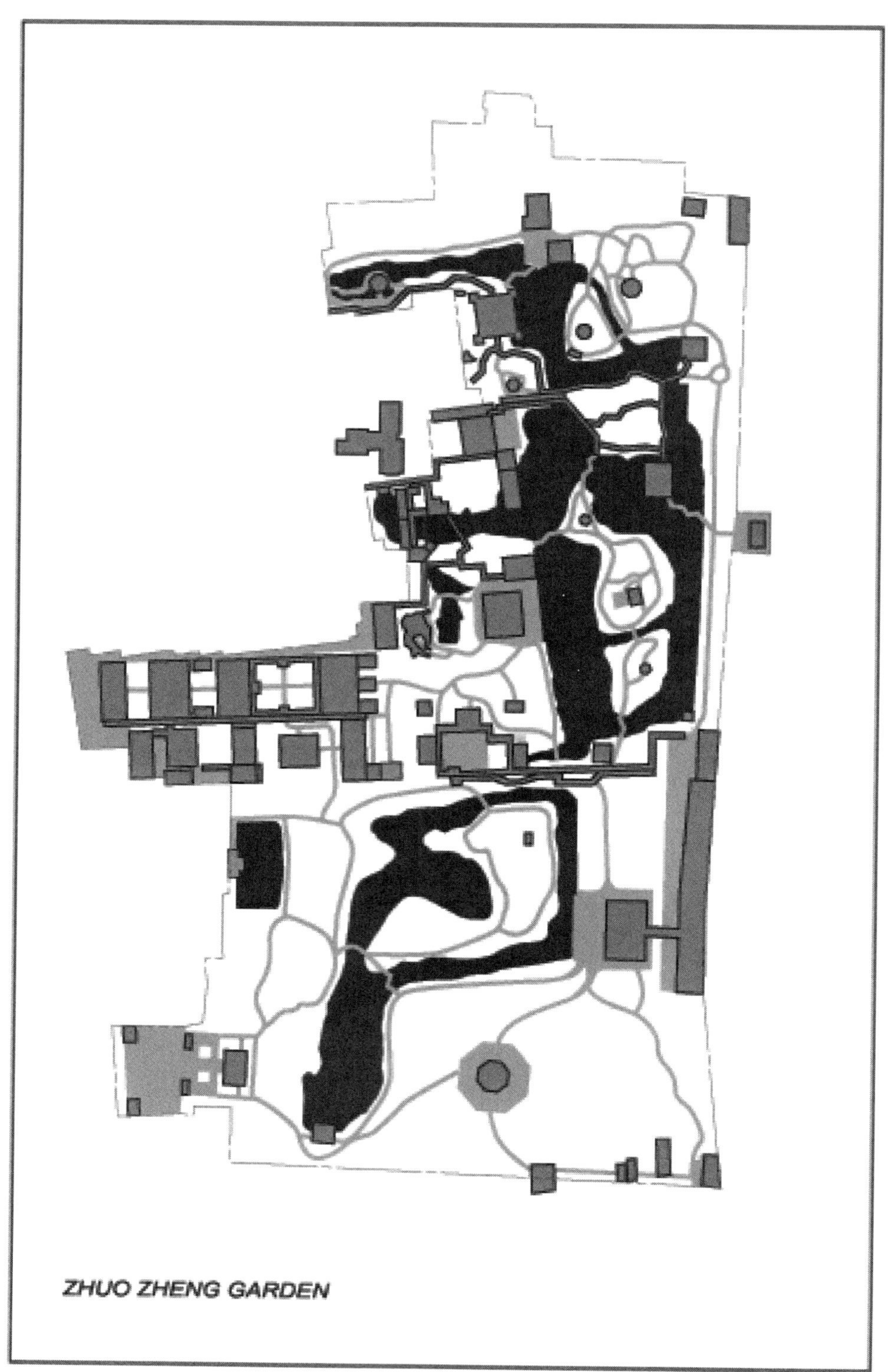

3-167

3-168

Today, after decades of renovation, one has to arrive early in the morning to avoid the thousands of tourists who crowd into this, the largest and most impressive of Suzhou's many wonderful water gardens.

REFERENCES

Gil, M. *Nature Perfected: The Story of the Garden.* Program Two: Ancient Spirits: China and Japan. Public Media Video. London, UK: Public Broadcasting System, 1995. Videocassette.

Wood, F. *Blue Guide: China.* New York: W. W. Norton & Co., 1992.

Zhou Zheng Garden. Suzhou, China: Ge Wu Xuan Publishing House, 1994.

REFERENCES

KEY REFERENCES

The following references were the sources for much of the material in this book. Highlighted sources represent components very useful (single asterisk) or essential (double asterisks) to any personal or office library for those seriously wishing to specialize in the design and creation of wetlands for environmental improvement or loss-replacement mitigation. Sources include books, symposia proceedings, reviews, and journals.

Wetland Management

Good, R. E., D. F. Whigham, R. L. Simpson, and C. G. Jackson. *Freshwater Wetland: Ecological Processes and Management Potential.* New York: Academic Press, 1978.

Kent, D. M. *Applied Wetlands Science and Technology.* Boca Raton, FL: Lewid, 1994.

*Kusler, J., and T. Opheim. *Our National Wetland Heritage: A Protection Guide.* Environmental Law Institute, 1996.

Leitch, J. A., and H. R. Ludwig. *Wetland Economics, 1989–1993: A Selected, Annotated Bibliography.* Westport, CT: Greenwood Press, 1995.

Mulamoottil, G., B. G. Warner, and E. A. McBean. *Wetlands: Environmental Gradients, Boundaries, and Buffers.* New York: Lewis, 1996.

**Ramsar Convention on Wetlands.* Parts 1–9. Land, Switzerland: The Ramsar Convention Bureau, 2000.

Wetland Creation

**Bartoldus, C. C., and E. W. Garbish. *Evaluation for Planned Wetlands (EPW): A Procedure for Assessing Wetland Functions and a Guide to Functional Design.* St. Michaels, MD: Environmental Concern, Inc., 1994.

Bavor, H. J., and D. S. Marshall, eds. "Wetland Systems in Water Pollution Control." *Water Science and Technology* 29(4) (1994): 1–338.

**Campbell, C. S., and M. H. Ogden. *Constructed Wetlands in the Sustainable Landscape.* New York: John Wiley & Sons, 1999.

*Davis, L. *A Handbook of Constructed Wetlands. A Guide to Creating Wetlands for: Agricultural Wastewater, Domestic Wastewater, Coal Mine Drainage.* Stormwater. U.S.E.P.A. Volumes 1–5. Washington, DC: Government Printing Office, 1995.

*Dunne, K. P., A. M. Rodrigo, and E. Samanns. *Engineering Guidelines for Wetland Plant Establishment and Subgrade Preparation.* Wetlands Research Program, U.S. Army Corps of Engineers. Washington, DC: Government Printing Office,1998.

*Environmental Protection Agency. *Constructed Wetlands for Wastewater Treatment and Wildlife Habitat: 17 Case Studies.* Washington, DC: Government Printing Office, 1993.

**France, R., M. Tucker, and L. Johnston. *Landscape Architecture of Created Wetlands: 17 Virtual Visual Tours.* Sheffield, VT: Green Frigate Books, 2002. CD-ROM.

*Hairston, A. J., ed. *Wetlands: An Approach to Improving Decision Making in Wetland Restoration and Creation.* Washington, DC: Island Press, 1992.

*Hamer, D. A. *Constructed Wetlands for Wastewater Treatment—Municipal, Industrial and Agricultural.* Chelsea, MI: Lewis, 1990.

Kadlec, R. H., and R. L. Knight. *Treatment Wetlands.* New York: CRC Lewis, 1996.

**Kusler, J. A., C. Ray, E. Zinecker, K. Savio, M. Klein, and S. Weaver. *Guidebook for Creating Wetland Interpretation Sites Including Wetlands & Ecotourism.* Berne, NY: Associate State Wetland Managers, 1998.

Kusler, J. A., and M. E. Kentula. *Wetland Creation and Restoration: The Status of the Science.* Washington, DC: Island Press, 1990.

*Mansell, D., L. Christl, R. Maher, A. Norman, N. Patterson, and T. Williams. *Temperate Wetlands Restoration Guidelines.* Barrie, Ontario: Ontario Ministry Natural Resources, 1998.

*Marble, A. D. *A Guide to Wetland Functional Design.* Ann Arbor, MI: Lewis, 1991.

Means, J. L., and R. E. Hinchee, eds. *Wetlands and Remediation.* Columbus, OH: Batelle Press, 1999.

*Moshiri, G. A. *Constructed Wetlands for Water Quality Improvement.* Ann Arbor, MI: Lewis, 1993.

Mulamoottil, E. A. McBean, and F. Rovers, eds. *Constructed Wetlands for the Treatment of Landfill Leachates.* Boca Raton, FL: Lewis, 1999.

Olson, R. K., ed. "The Role of Treated and Natural Wetlands in Controlling Nonpoint Source Pollution." *Ecological Engineering* 1 (1992): 1–170.

Olson, R. K., ed. *Created and Natural Wetlands for Controlling Nonpoint Source Pollution.* Boca Raton, FL: CRC Press, 1993.

*Pierce, G. J. *Planning Hydrology for Constructed Wetlands.* West Clarksville, NY: Wetlands Training Institute/Southern Tier Consulting, 1993.

Reed, S. C., R. W. Crites, and E. J. Middlebrooks. *Natural Systems for Waste Management and Treatment.* New York: McGraw-Hill, 1998.

Reddy, K. R., and P. M. Gale, eds. "Wetland Processes and Water Quality: A Symposium." *Journal of Environmental Quality* 23 (1994): 875–1625.

Reddy, K. R., and W. H. Smith. *Aquatic Plants for Water Treatment and Resource Recovery.* Orlando, FL: Magnolia, 1987.

*Southern Tier Consulting. *Wetland Construction and Restoration.* West Clarksville, NY: Wetland Training Institute, 2000.

Thunhorst, G. A. *Wetland Planting Guide for the Northeastern United States: Plants for Wetland Creation, Restoration, and Enhancement.* St. Michaels, MD: Environmental Concern Inc., 1993.

United States Department of Agriculture. "Wetland Restoration, Enhancement, or Creation." In *Engineering Field Handbook.* U.S.D.A. Soil Conservency Service. Washington, DC: Government Printing Office.

Vymazal, J., H. Brix, P. F. Cooper, M. B. Green, and R. Haberl, eds. *Constructed Wetlands for Wastewater Treatment in Europe.* London: Backhuys, 1998.

**The Wetland Journal.* St. Michaels, MD: Environmental Concern Inc., 1997–2000.

Water Gardens

Burrell, C. C., ed. *The Natural Water Garden.* Brooklyn, NY: Brooklyn Botanic Garden, 1997.

*Glattstein, J. *Waterscaping: Plants and Ideas for Natural and Created Water Gardens.* London, UK: Garden Way, 1994.

Jansen, A. *Success With Plants for Your Garden Pond.* Putney, England: Cavendish Books, Merehurst Ltd., 1994.

Stadelmann, P. *Success With Your Garden Pond.* Putney, England: Cavendish Books, Merehurst Ltd., 1989.

Stein, S. *Water Gardening.* Abingdon, England: Transedition Books,1994.

Thomas, C. B. *Water Gardens: How to Plan and Plant a Backyard Pond.* Boston, MA: Houghton Mifflin, 1997.

Water Gardens. Menlo Park, CA: Sunset Books Inc., 1997.

Wetland History, Loss, Development Pressures, and Mitigation

Hey, D. L., and N. S. Philippi. *A Case for Wetland Restoration.* New York: John Wiley & Sons, 2000.

**Salvesen, D. Wetlands: *Mitigating and Regulating Development Impacts.* Washington, DC: The Urban Land Institute, 1994.

Vilesisis, A. *Discovering the Unknown Landscape: A History of America's Wetlands.* Washington, DC: Island Press, 1997.

Wetland Mitigation. *Ecological Applications* 6 (1996): 33–163.

Wetland Culture

*Giblet, R. *Postmodern Wetlands: Culture, History, Ecology.* Edinburgh: Edinburgh University Press, 1996.

**Hurd, B. *Stirring the Mud: On Swamps, Bogs, and Human Imagination.* Boston, MA: Beacon Press, 2001.

*Wilson, S., and T. Moritz. *The Sierra Club Wetlands Reader: A Literary Companion.* San Francisco, CA: Sierra Club Books, 1996.

Watershed-scale Planning

**Dramstad, W. E., J. O. Olson, and R. T. T. Forman. *Landscape Ecology Principles in Landscape Architecture and Land-use Planning.* Washington, DC: Island Press, 1996.

*Environmental Protection Agency. *Top 10 Watershed Lessons Learned.* Washington, DC: Government Printing Office, 1997.

**France, R., ed. *Handbook of Water Sensitive Planning and Design.* Boca Raton, FL: CRC/Lewis Publishers. In Press, 2002.

Kusler, J. A., D. E. Willard, and H. C. Hull, Jr. *Wetlands and Watershed Management: Science Applications and Public Policy.* Berne, NY: Associated State Wetland Managers, 1997.

Lyon, J. G., and J. McCarthy, eds. *Wetland and Environmental Applications of GIS.* Boca Raton, FL: Lewis, 1995.

"Wetland Mitigation in a Landscape Context." *Environmental Management* 12 (1988): 25: 130–80.

ADDITIONAL REFERENCES

The following references are for individual papers or reports not covered in the books or special journal issues listed previously. All sources are general reviews.

Cole, S. "The Emergence of Treatment Wetlands." *Environmental Science & Technology* 223 (1998): 218–223.

Ellis, J. B., R. B. Shutes, D. M. Revitt, and T. T. Zhang. "Use of Macrophytes for Pollution Treatment in Urban Wetlands." *Resources, Conservation and Recycling* 11 (1994): 1–12.

Environmental Protection Agency, Gulf of Mexico Program. *Constructed Wetlands and Wastewater Management for Confined Animal Feeding Operations.* Washington, DC: Government Printing Office, 1996.

Hamilton, H., P. G. Nix, and A. Sobolewski. "An Overview of Constructed Wetlands as Alternatives to Conventional Waste Treatment Systems." *Water Pollution Research Journal of Canada* 28 (1993): 529–548.

Jarman, N. M., R. A. Dobberteen, B. Windmiller, and P. R. Lelito. "Evaluation of Created Freshwater Wetlands in Massachusetts." *Restoration & Management Notes* 9 (1991): 26–29.

Malakoff, D. "Restored Wetlands Flunk Real-world Test." *Science* 280 (1998): 371–372.

Mitsch, W. J. "Combining Ecosystem and Landscape Approaches to Great Lakes Wetlands." *Journal of Great Lakes Research* 18 (1992): 552–570.

Rosen, D. K. "Wetland Restoration Step by Step." *Water Gardening Magazine* May/June (1997): 14–21.

Shutes, R., D. M. Revitt, A. S. Mungur, and L. N. Scholes. "Design of Wetland Systems for the Treatment of Urban Runoff." *Water Quality* 25 (1997): 35–38.

Taylor, M. *Constructed Wetlands for Stormwater Management: A Review.* Toronto, Ontario: Ontario Ministry of the Environment, 1992.

Tennessee Valley Authority. General Design, *Construction, and Operation Guidelines: Constructed Wetlands Wastewater Treatment Systems for Small Users Including Individual Residences.* Nashville, TN: Tennessee Valley Authority, 1991.

Williams, T. "What Good is a Wetland?" *Audobon Magazine* November/December (1996): 42–53.

GLOSSARY

Although I have made a conscious effort to avoid using technical jargon, this was not always possible. The following terms—whose meaning may not be immediately understood—are left undefined in the body of the text.

BIOLOGY

Coldwater fisheries: contain commercially important fishes such as salmon or trout.

Emergent vegetation: typical tall marsh plants, such as cattails, that stand above the height of the water.

Invasive exotics: non-native species characterized by a rapid rate of colonization

ECOLOGY

Ecotones: spatially fuzzy regions located between different ecosystems.

functional attributes: concerned with dynamic variables such as population growth rates or system productivity.

guild analysis: groups of organisms having similar ecological or evolutionary traits.

riparian: special type of ecotone situated between terrestrial and aquatic environments.

succession: natural process of serial replacement of some species by others during system maturation.

trophic structure: description of foodweb relationships within an ecological community.

GEOCHEMISTRY

Budgets: mass balance calculation of all inputs and losses of an element within an ecosystem.

Mass loading rates: rate input of an element (amount per unit time) from all sources.

HYDROLOGY

Channelization: the ditching and straightening of river meanders.

Daylighting: the process of opening up once-buried urban streams to the light of day.

Dissipater: a rock structure at the inlet end of a wetland to slow down and dissipate inflowing surges of stormwater.

Drawdown: the process of deliberately dewatering a wetland.

Freeboard: the vertical distance from the surface of the water to the top of the control berm or weir.

Groundwater recharge: the slow infiltration of surface runoff through the soil and into the subsurface reservoir of water, where it serves as a supply source for streams.

Weirs: hydraulic control structures that regulate the flow of water

STATISTICS

Multiple regression models: a mathematical expression of the strength of the relationship between a variable of interest to a suite of other, potentially determining, variables.

WETLANDS

Internal microtopography: structures, such as berms, used to direct pathways of water flow.

Treatment cells: partially self-contained regions within a wetland that interrelate through water flow.

Order: the spatial location of natural wetlands on the landscape, from first-order systems situated in the headwaters of watershed to higher-order systems situated further downstream.

Polishing: the final stage of water purification in a treatment wetland, following removal of much of the contaminant load in preceding treatment cells.

INDEX